MW00677504

Be prepared...
To learn...
To succeed...

Get **REA**dy. It all starts here.
REA's preparation for the CRCT is **fully aligned**
with Georgia Performance Standards.

Visit us online at
www.rea.com

READY, SET, GO!®

Georgia
CRCT
2nd Edition
Grade 8
Mathematics

 With REA's TestWare® on CD-ROM

Stephen A. Reiss, M.B.A.
The Math Magician
and
Reiss SAT Seminars

Research & Education Association
Visit our website at
www.rea.com

This book aligns with Georgia Performance Standards.
For further information, visit the Georgia
Department of Education website at
http://www.georgiastandards.org.

Research & Education Association
61 Ethel Road West
Piscataway, New Jersey 08854
E-mail: info@rea.com

Ready, Set, Go! ®
**Georgia CRCT Mathematics, Grade 8
with TestWare® on CD-ROM**

Copyright © 2010 by Research & Education Association, Inc.
Prior edition copyright © 2006 All rights reserved. No part of this
book may be reproduced in any form without permission of the
publisher.

Printed in the United States of America

Library of Congress Control Number 2009935261

ISBN-13: 978-0-7386-0684-2

ISBN-10: 0-7386-0684-7

Windows® is a registered trademark of Microsoft Corporation.

A10-0101

TABLE OF CONTENTS

About Research & Education Association

Founded in 1959, Research & Education Association is dedicated to publishing the finest and most effective educational materials—including software, study guides, and test preps—for students in elementary school, middle school, high school, college, graduate school, and beyond. Today, REA's wide-ranging catalog is a leading resource for teachers, students, and professionals.

We invite you to visit us at *www.rea.com* to find out how "REA is making the world smarter."

Acknowledgments

We would like to thank Larry B. King, Vice President, Editorial, for his editorial direction; Pam Weston, Vice President, Publishing, for setting the quality standards for production integrity and managing the publication to completion; John Cording, Vice President, Technology for coordinating the design and development of REA's TestWare®; Alice Leonard, Senior Editor, for project management; Diane Goldschmidt, Senior Editor, for post-production quality assurance; Mel Friedman, Senior Editor, Mathematics, for reviewing the work; Heena Patel, Technology Project Manager, for software testing; Christine Saul, Senior Graphic Artist, for cover design; Sandra Rush, for copyediting; and Macmillan Publishing Solutions for typesetting the book.

SUCCEEDING ON THE CRCT— GRADE 8 MATHEMATICS

ABOUT THIS BOOK AND TestWare®

This book and the accompanying TestWare® provide excellent preparation for the Georgia Criterion-Referenced Competency Tests (CRCT)—Grade 8 Mathematics. Inside you will find exercises designed to provide students with the instruction, practice, and strategies needed to do well on this achievement test.

This book is divided into several sections:

- The first section contains eight chapters, each of which is specifically designed to match an objective, as stated by the Georgia Department of Education. Each chapter consists of illustrative examples and an in-depth review of all test topics. The material is fully aligned with the domains and standards created by the Georgia Department of Education.

- The second section contains two practice tests. Each practice test consists of 70 questions with detailed explanations. Each question is identified by the domain and standard that it measures.

- The third section contains a student score sheet that identifies the topic associated with each question of both practice tests. This score sheet can be used to help identify a student's strengths and weaknesses

HOW TO USE THIS BOOK AND TestWare®

FOR STUDENTS: You'll find that our practice tests are very much like the actual CRCT you'll encounter on test day. The best way to prepare for a test is to practice, so we've included drills with answers throughout the book, and our two practice tests include detailed answers. Taking the practice tests on your computer will help you with the timing on the CRCT test.

FOR PARENTS: Georgia has created grade-appropriate performance standards and core curriculum requirements that are listed in the table in this introduction. Students need to meet these objectives as measured by the CRCT. Our book will help your child review for the CRCT and prepare for the Mathematics exam. It includes review sections, drills, and two practice tests complete with explanations to help your child focus on the areas he/she needs to work on to help master the test.

FOR TEACHERS: No doubt, you are already familiar with the CRCT and its format. Begin by assigning students the diagnostic tests. An answer key and detailed explanations follow the diagnostic tests. Then work through each of the lessons in succession. When students have completed the subject review, they should move on to the practice tests. Answers and answer explanations follow the practice tests.

WHY STUDENTS ARE REQUIRED TO TAKE THE CRCT

CRCT measures the extent to which students are meeting the Georgia Performance Standards (GPS). Georgia teachers and curriculum experts developed the CRCT in cooperation with the Georgia Department of Education. The Mathematics test is given to students in grades 1 through 8.

WHAT'S ON THE CRCT

The mathematics test in Grade 8 is given in two 70-minute sessions, with a 10-minute break between sessions. Each session consists of 35 questions for a total of 70 questions.

All questions on the CRCT are multiple-choice. Multiple-choice questions are based on the Georgia Performance Standards, which are listed on the following pages.

The approximate number of questions per domain is as follows.

Domain 1: Numbers and Operations	15
Domain 2: Geometry	9
Domain 3: Algebra	35
Domain 4: Date Analysis and Probability	11
Total number of questions:	**70**

In addition, a topic entitled *Mathematical Process Skills* is integrated across all four domains. Each Mathematical Process Skills question can be identified with one of the basic four domains. This type of question refers to students' ability to apply content knowledge to problem solving. This ability also applies to:

a. reasoning and evaluating mathematical arguments

b. making connections among mathematical ideas

c. representing mathematical ideas in multiple ways

GEORGIA DEPARTMENT OF EDUCATION— LIST OF DOMAINS FOR CRCT 8TH GRADE MATHEMATICS

Domain 1: Numbers and Operations

M8N1 Students will understand different representations of numbers including square roots, exponents, and scientific notation.

Elements:

a. Find square roots of perfect squares.

b. Recognize the (positive) square root of a number as a length of a side of a square with a given area.

c. Recognize square roots as points and as lengths on a number line.

d. Understand that the square root of 0 is 0 and that every positive number has two square roots that are opposite in sign.

e. Recognize and use the radical symbol to denote the positive square root of a positive number.

f. Estimate square roots of positive numbers.

g. Simplify, add, subtract, multiply, and divide expressions containing square roots.

h. Distinguish between rational and irrational numbers.

i. Simplify expressions containing integer exponents.

j. Express and use numbers in scientific notation.

k. Use appropriate technologies to solve problems involving square roots, exponents, and scientific notation.

Approximately 15 Questions on Actual Test

Domain 2: Geometry

M8G1 Students will understand and apply the properties of parallel and perpendicular lines and understand the meaning of congruence.

Elements:

a. Investigate characteristics of parallel and perpendicular lines both algebraically and geometrically.

b. Apply properties of angle pairs formed by parallel lines cut by a transversal.

c. Understand the properties of the ratio of segments of parallel lines cut by one or more transversals.

d. Understand the meaning of congruence: that all corresponding angles are congruent and all corresponding sides are congruent.

M8G2 Students will understand and use the Pythagorean theorem.

Elements:

a. Apply properties of right triangles, including the Pythagorean theorem.

b. Recognize and interpret the Pythagorean theorem as a statement about areas of squares on the sides of a right triangle.

Approximately 9 Questions on Actual Test

Domain 3: Algebra

M8A1 Students will use algebra to represent, analyze, and solve problems.

Elements:

a. Represent a given situation using algebraic expressions or equations in one variable.

b. Simplify and evaluate algebraic expressions.

c. Solve algebraic equations in one variable, including equations involving absolute values.

d. Solve equations involving several variables for one variable in terms of the others.

e. Interpret solutions in problem contexts.

M8A2 Students will understand and graph inequalities in one variable.

Elements:

a. Represent a given situation using an inequality in one variable.

b. Use the properties of inequality to solve inequalities.

c. Graph the solution of an inequality on a number line.

d. Interpret solutions in problem contexts.

M8A3 Students will understand relations and linear functions.

Elements:

    a. Recognize a relation as a correspondence between varying quantities.

    b. Recognize a function as a correspondence between inputs and outputs where the output for each input must be unique.

    c. Distinguish between relations that are functions and those that are not functions.

    d. Recognize functions in a variety of representations and a variety of contexts.

    e. Use tables to describe sequences recursively and with a formula in closed form.

    f. Understand and recognize arithmetic sequences as linear functions with whole-number input values.

    g. Interpret the constant difference in an arithmetic sequence as the slope of the associated linear function.

    h. Identify relations and functions as linear or nonlinear.

    i. Translate among verbal, tabular, graphic, and algebraic representations of functions.

M8A4 Students will graph and analyze graphs of linear equations and inequalities.

Elements:

    a. Interpret slope as a rate of change.

    b. Determine the meaning of the slope and y-intercept in a given situation.

    c. Graph equations of the form $y = mx + b$.

    d. Graph equations of the form $ax + by = c$.

    e. Graph the solution set of a linear inequality, identifying whether the solution set is an open or a closed half-plane.

    f. Determine the equation of a line given a graph, numerical information that defines the line, or a context involving a linear relationship.

    g. Solve problems involving linear relationships.

M8A5 Students will understand systems of linear equations and inequalities and use them to solve problems.

Elements:

    a. Given a problem context, write an appropriate system of linear equations or inequalities.

b. Solve systems of equations graphically and algebraically, using technology as appropriate.

c. Graph the solution set of a system of linear inequalities in two variables.

d. Interpret solutions in problem contexts.

Approximately 35 Questions on Actual Test

Domain 4: Data Analysis and Probability

M8D1 Students will apply basic concepts of set theory.

Elements:

a. Demonstrate relationships among sets through use of Venn diagrams.

b. Determine subsets, complements, intersection, and union of sets.

c. Use set notation to denote elements of a set.

M8D2 Students will determine the number of outcomes related to a given event.

Elements:

a. Use tree diagrams to find the number of outcomes.

b. Apply the addition and multiplication principles of counting.

M8D3 Students will use the basic laws of probability.

Elements:

a. Find the probability of simple independent events.

b. Find the probability of compound independent events.

M8D4 Students will organize, interpret, and make inferences from statistical data.

Elements:

a. Gather data that can be modeled with a linear function.

b. Estimate and determine a line of best fit from a scatter plot

Approximately 11 Questions on Actual Test

Mathematical Process Skills are integrated across the four domains.

Following are the associated standards.

M8P1 Students will solve problems (using appropriate technology).

Elements:

    a. Build new mathematical knowledge through problem solving.

    b. Solve problems that arise in mathematics and in other contexts.

    c. Apply and adapt a variety of appropriate strategies to solve problems.

    d. Monitor and reflect on the process of mathematical problem solving.

M8P2 Students will reason and evaluate mathematical arguments.

Elements:

    a. Recognize reasoning and proof as fundamental aspects of mathematics.

    b. Make and investigate mathematical conjectures.

    c. Develop and evaluate mathematical arguments and proofs.

    d. Select and use various types of reasoning and methods of proof.

M8P3 Students will communicate mathematically.

Elements:

    a. Organize and consolidate their mathematical thinking through communication.

    b. Communicate their mathematical thinking coherently and clearly to peers, teachers, and others.

    c. Analyze and evaluate the mathematical thinking and strategies of others.

    d. Use the language of mathematics to express mathematical ideas precisely.

M8P4 Students will make connections among mathematical ideas and to other disciplines.

Elements:

    a. Recognize and use connections among mathematical ideas.

    b. Understand how mathematical ideas interconnect and build on one another to produce a coherent whole.

c. Recognize and apply mathematics in contexts outside of mathematics.

M8P5 Students will represent mathematics in multiple ways.

Elements:

a. Create and use representations to organize, record, and communicate mathematical ideas.

b. Select, apply, and translate among mathematical representations to solve problems.

c. Use representations to model and interpret physical, social, and mathematical phenomena

TIPS FOR THE STUDENT

Students can do plenty of things before and during the actual test to improve their performance. The good thing is that most of the tips described in the following pages are easy!

Preparing for the Test

Test Anxiety

Do you get nervous when your teacher talks about taking a test? A certain amount of anxiety is normal and it actually may help you prepare better for the test by getting you motivated. But too much anxiety is a bad thing and may keep you from properly preparing for the test. Here are some things to consider that may help relieve test anxiety:

- Share how you are feeling with your parents and your teachers. They may have ways of helping you deal with how you are feeling.

- Keep on top of your game. Are you behind in your homework and class assignments? A lot of your classwork-related anxiety and stress will simply go away if you keep up with your homework assignments and classwork. And then you can focus on the test with a clearer mind.

- Relax. Take a deep breath or two. You should do this especially if you get anxious while taking the test.

Study Tips & Taking the Test

- **Learn the Test's Format.** Don't be surprised. By taking a practice test ahead of time you'll know what the test looks like, how much time you will have, how many questions there are, and what kinds of questions are going to appear on it. Knowing ahead of time is much better than being surprised.

- **Read the Entire Question.** Pay attention to what kind of answer a question or word Problem is looking for. Reread the question if it does not make sense to you, and try to note the parts of the question needed for figuring out the right answer.

- **Read All the Answers.** On a multiple-choice test, the right answer could also be the last answer. You won't know unless you read all the possible answers to a question.

- **It's Not a Guessing Game.** If you don't know the answer to a question, don't make an uneducated guess. And don't randomly pick just any answer either.

As you read over each possible answer to a question, note any answers which are obviously wrong. Each obviously wrong answer you identify and eliminate greatly improves your chances at selecting the right answer.

- **Don't Get Stuck on Questions.** Don't spend too much time on any one question. Doing this takes away time from the other questions. Work on the easier questions first. Skip the really hard questions and come back to them if there is still enough time.

- **Accuracy Counts.** Make sure you record your answer in the correct space on your answer sheet. Fixing mistakes only takes time away from you.

- **Finished Early?** Use this time wisely and double-check your answers.

Sound Advice for Test Day

The Night Before. Getting a good night's rest keeps your mind sharp and focused.

The Morning of the Test. Have a good breakfast. Dress in comfortable clothes. Keep in mind that you don't want to be too hot or too cold while taking the test. Get to school on time. Give yourself time to gather your thoughts and calm down before the test begins.

Three Steps for Taking the Test

1) **Read.** Read the entire question and then read all the possible answers.

2) **Answer.** Answer the easier questions first and then go back to the more difficult questions.

3) **Double-Check.** Go back and check your work if time permits.

Tips for Parents

- Encourage your child to take responsibility for homework and class assignments. Help your child create a study schedule. Mark the test's date on a family calendar as a reminder for both of you.

- Talk to your child's teachers. Ask them for progress reports on an ongoing basis.

- Commend your child's study and test successes. Praise your child for successfully following a study schedule, for doing homework, and for work done well.

- Test Anxiety. Your child may experience nervousness or anxiety about the test. You may even be anxious, too. Here are some helpful tips on dealing with a child's test anxiety:

- Talk about the test openly and positively with your child. An ongoing dialogue not only can relieve your child's anxieties but also serves as a progress report of how your child feels about the test.

- Form realistic expectations of your child's testing abilities.

- Be a "Test Cheerleader." Your encouragement to do his or her best on the test can alleviate your child's test anxiety.

Arithmetic Diagnostic Test— Georgia CRCT

1. One tenth is what fraction of three fourths?

    (A) $\dfrac{3}{40}$

    (B) $\dfrac{1}{8}$

    (C) $\dfrac{2}{15}$

    (D) $\dfrac{15}{2}$

2. What percent of 260 is 13?

    (A) 20%

    (B) 5%

    (C) 0.5%

    (D) 0.05%

3. $4\dfrac{1}{3} - 1\dfrac{5}{6} = $ _____.

    (A) $2\dfrac{1}{6}$

    (B) $2\dfrac{1}{2}$

    (C) $3\dfrac{1}{2}$

    (D) $3\dfrac{2}{3}$

GO ON

4. Which of the following has the smallest value?

   (A) $\dfrac{0.1}{2}$

   (B) $\dfrac{1}{0.2}$

   (C) $\dfrac{0.2}{1}$

   (D) $\dfrac{0.2}{0.1}$

5. What is the value of $10 - 5[2^3 + 27 \div 3 - 2(8 - 10)]$?

   (A) -95

   (B) -55

   (C) 65

   (D) 105

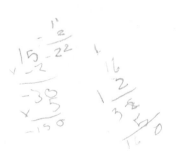

6. What is the value of $\dfrac{2^{-4} + 2^{-1}}{2^{-3}}$?

   (A) $\dfrac{1}{4}$

   (B) $\dfrac{1}{2}$

   (C) 1

   (D) $\dfrac{9}{2}$

GO ON

7. $\frac{2}{3}+\frac{5}{9}=$ _____.

    (A) $\frac{7}{12}$

    (B) $\frac{11}{9}$

    (C) $\frac{7}{3}$

    (D) $\frac{11}{3}$

$$\frac{6}{9}+\frac{5}{9}=\frac{11}{9}=1\frac{2}{9}$$

8. What is the decimal equivalent of 125.937%?

    (A) 1.25937

    (B) 12.5937

    (C) 1259.37

    (D) 12,593.7

9. What is the reduced ratio of 8 feet to 28 inches?

    (A) $\frac{2}{7}$

    (B) $\frac{6}{7}$

    (C) $\frac{24}{7}$

    (D) $\frac{7}{2}$

10. Which of the following is the most economical price of an item?

    (A)   10 ounces for 16 cents

    (B)   2 ounces for 3 cents

    (C)   4 ounces for 7 cents

    (D)   20 ounces for 34 cents

11. As a reduced improper fraction, $4\dfrac{5}{6} = $ _____.

    (A) $\dfrac{5}{24}$

    (B) $\dfrac{3}{2}$

    (C) $\dfrac{13}{3}$

    (D) $\dfrac{29}{6}$

12. $4\% \times 4\% = $ _____.

    (A)   0.0016%

    (B)   0.16%

    (C)   1.6%

    (D)   16%

GO ON

13. Which of the following numbers is *not* between $0.\overline{85}$ and $0.\overline{86}$?

(A) $0.\overline{851}$

(B) $0.\overline{861}$

(C) 0.859

(D) 0.861

14. What is the decimal equivalent of $\dfrac{7}{8}$?

(A) 0.666

(B) 0.777

(C) 0.875

(D) 1.142

15. What is the simplified form of $6\sqrt{7} + 4\sqrt{7} - \sqrt{5} + \sqrt{7}$?

(A) $10\sqrt{7} + \sqrt{2}$

(B) $11\sqrt{7} - \sqrt{5}$

(C) $9\sqrt{7} - \sqrt{5}$

(D) 44

16. What is the simplified form of $\sqrt{75} - 3\sqrt{48}$?

(A) $7\sqrt{3}$

(B) $3\sqrt{3}$

(C) $-3\sqrt{3}$

(D) $-7\sqrt{3}$

GO ON

17. Fifteen percent of what number is 60?

    (A) 9
    (B) 45
    (C) 400
    (D) 900

18. The digit 4 in the number 17.3541 is in which place?

    (A) Tenths
    (B) Hundredths
    (C) Thousandths
    (D) Ten-thousandths

19. What is the value of $3^4 - 2^5 - 1^6$?

    (A) 13
    (B) 30
    (C) 43
    (D) 48

GO ON

20. Matilda's car gets 34 miles per gallon and Naomi's car gets only 8 miles per gallon. In traveling from city A to city B, each car uses a whole number of gallons of gasoline. Which of the following could represent the distance between cities A and B?

(A) 42 miles

(B) 74 miles

(C) 106 miles

(D) 136 miles

21. Joel earns $6.75 per hour at his part-time job. What is the minimum number of hours he must work in order to earn at least $150?

(A) 20

(B) 22

(C) 23

(D) 24

22. Mary has $29\frac{1}{2}$ yards of material available to make uniforms. Each uniform requires $\frac{3}{4}$ yard of material. How many uniforms can she make and how much material will she have left?

(A) 39 uniforms with $\frac{1}{3}$ yard left over

(B) 39 uniforms with $\frac{1}{4}$ yard left over

(C) 27 uniforms with $\frac{1}{2}$ yard left over

(D) 27 uniforms with $\frac{1}{3}$ yard left over

GO ON

23. Of the students enrolled in remedial mathematics at a college, 20% failed this course. If 600 students received passing scores, how many students enrolled in this course?

    (A) 3000
    (B) 1800
    (C) 1350
    (D)   750

24. If a baseball cap is regularly priced at $16.00, how much will it cost during a 15%-off sale?

    (A) $12.00
    (B) $12.75
    (C) $13.60
    (D) $14.20

25. On the first day of December, the highest temperature for a certain town was 32°. On January 1st, the highest temperature had dropped 38 degrees. What was the highest temperature on January 1st?

    (A) −6°
    (B) −2°
    (C) 2°
    (D) 6°

Algebra Diagnostic Test— Georgia CRCT

1. What is the expression for B in the equation $A = (h/2)(B + b)$?

    (A) $(2A - b)/h$

    (B) $2h/A - b$

    (C) $2A - bh$

    (D) $2A/h - b$

2. If x is an integer, which of the following *must* represent a perfect square?

    (A) $x^2 + 2x + 1$

    (B) $x^2 + 1$

    (C) $x^2 + x$

    (D) $x^2 + 2x + 4$

3. What is the value of x in the equation $\sqrt{5x - 4} - 5 = -1$?

    (A) 2

    (B) 4

    (C) 5

    (D) No solution

4. If n is any integer, which of the following *must* represent an odd integer?

    (A) $n + 1$

    (B) $2n$

    (C) $2n + 3$

    (D) $3n$

5. If $x - y = 9$, then $3x - 3y - 1 = $ _____.

    (A) 24

    (B) 25

    (C) 26

    (D) 27

6. If $\frac{3}{2}x = 5$, what is the value of $x + \frac{2}{3}$?

    (A) $\frac{10}{3}$

    (B) 4

    (C) $\frac{15}{2}$

    (D) $\frac{49}{6}$

7. What is the value of x in the equation $\frac{x}{2/3} = \frac{30}{5}$?

    (A) 4

    (B) 5

    (C) 6

    (D) 7

8. If $f(x) = \frac{45}{x}$, which of the following ordered pairs lies on the graph of $f(x)$?

    (A) $(45, 0)$

    (B) $(5, 9)$

    (C) $(9, 6)$

    (D) $(40, 5)$

9. Which of the following is equivalent to $6x - 4 - (5 - x)$?

    (A) $5x + 9$

    (B) $5x - 9$

    (C) $7x + 9$

    (D) $7x - 9$

10. Which of the following describes the statement "Twenty three more than x is equal to x subtracted from fifty five?"

    (A) $23x = 55 - x$

    (B) $23x = x - 55$

    (C) $x + 23 = 55 - x$

    (D) $x + 23 = x - 55$

11. What is the value of $5x - 2y$ if $x = 2$ and $y = -3$?

    (A) -16

    (B) -4

    (C) 16

    (D) 4

12. What is the value of x in the equation $4(x + 3) + 6x = 22$?

    (A) -3

    (B) -1

    (C) 1

    (D) 3

GO ON

13. In the following table, y is a linear function of x.

x	2	3	5	8
y	1	5	13	—

What is the missing y value?

(A) 16

(B) 19

(C) 22

(D) 25

14. The ratio of boys to girls in a school auditorium is 7 to 8. If there are 320 girls which proportion can be used to find the number of boys, b?

(A) $\dfrac{7}{8} = \dfrac{320}{b}$

(B) $\dfrac{7}{8} = \dfrac{b}{320}$

(C) $\dfrac{8}{15} = \dfrac{b}{320}$

(D) $\dfrac{7}{15} = \dfrac{b}{320}$

15. Which inequality matches the following graph?

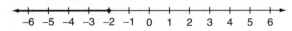

(A) $x \leq -2$

(B) $x < -2$

(C) $x \geq -2$

(D) $x > -2$

GO ON

16. Which of the following has no solution for x?

    (A) $|x-4|+8=4$

    (B) $|x-2|+1=1$

    (C) $|x-3|-5=-3$

    (D) $|x+6|=|x-6|$

17. What is the value of y in the following system of equations?

$$x - y = 19$$
$$3x + y = -7$$

    (A) -20

    (B) -16

    (C) 3

    (D) 13

18. What is the slope of a line that contains the points $(-4, -3)$ and $(-6, 7)$?

    (A) -2

    (B) -5

    (C) 2

    (D) 5

19. What is the y-intercept of the graph of the line whose equation is $3x - 5y = -15$?

    (A) $(-5, 0)$

    (B) $(5, 0)$

    (C) $(0, -3)$

    (D) $(0, 3)$

GO ON

20. One square has a side of length x, while a second square has a side of length $x + 2$. Which of the following is an expression for the sum of the areas of the two squares?

   (A) $2x^2 + 4x + 4$

   (B) $x^2 + 2$

   (C) $2x^2 + 4$

   (D) $2x^2 + 2x + 2$

21. For which one of the following is y *not* a function of x?

   (A) $9x + y = 12$

   (B) $5x^2 + y = 10$

   (C) $x + y^2 = 20$

   (D) $\sqrt{x} + y = 100$

22. Peter bought n compact disks for m dollars. How many compact disks could he buy for q dollars?

   (A) $\dfrac{m}{nq}$

   (B) $\dfrac{nq}{m}$

   (C) $\dfrac{q}{mn}$

   (D) $\dfrac{mn}{q}$

23. Emily receives a weekly salary of $400 plus a 12% commission of the total volume of sales she makes. If she wishes to make a total weekly salary of $700, what must be her dollar volume?

   (A) $3600

   (B) $3000

   (C) $2800

   (D) $2500

24. A truck contains 150 packages, some weighing 1 kg each and some weighing 3 kg each. If the total weight of all the packages is 264 kg, how many 3-kg packages are there?

    (A)   38

    (B)   57

    (C)   93

    (D)   112

25. Jake sold two-thirds of his pencils for 20 cents each. If he has 7 pencils left, how much money did he collect for the pencils he sold?

    (A) $1.40

    (B) $1.80

    (C) $2.10

    (D) $2.80

Geometry Diagnostic Test— Georgia CRCT

1. Which of the following has no dimension?

    (A) line
    (B) point
    (C) ray
    (D) triangle

2. The measures of two angles in a triangle are 60° and 12°. What is the measure of the third angle?

    (A) 18°
    (B) 48°
    (C) 72°
    (D) 108°

3. If the angle between two lines is 90°, then the two lines are _____ to each other.

    (A) parallel
    (B) adjacent
    (C) perpendicular
    (D) skew

GO ON

For 4 and 5, use the following diagram, in which lines l_1 and l_2 are parallel to each other. Line l_3 intersects both l_1 and l_2.

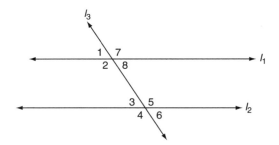

4. Which one of the following angles does *not* have the same measure as $\angle 3$?

  (A) $\angle 1$
  (B) $\angle 6$
  (C) $\angle 7$
  (D) $\angle 8$

5. Which one of the following pairs of angles are called alternate interior angles?

  (A) $\angle 2$ and $\angle 5$
  (B) $\angle 6$ and $\angle 8$
  (C) $\angle 3$ and $\angle 4$
  (D) $\angle 3$ and $\angle 5$

6. If an angle is obtuse, which one of the following could be its measure?

  (A) 88°
  (B) 155°
  (C) 180°
  (D) 264°

GO ON

7. In which one of the following diagrams could you conclude that ∠1 and ∠2 are complementary angles?

(A)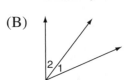

(B)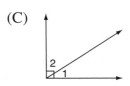

(C)

(D)

8. Which of the following is the correct symbol for this geometric figure?

```
    A                    B
  •————————————————————•
```

(A) $\overrightarrow{BA}$

(B) $\overrightarrow{AB}$

(C) $\overleftarrow{AB}$

(D) $\overline{AB}$

9. Two legs of a right triangle are 6 inches and 8 inches. How many inches long is the hypotenuse?

(A) 9

(B) 10

(C) 11

(D) 12

GO ON

10. The perimeter of a triangle is 40. If the length of one side is 12 and the other two sides are equal in length, what is the length of either of the other two sides?

   (A) 8
   (B) 9
   (C) 13
   (D) 14

11. The area of a square is 81. What is the perimeter of this square?

   (A) 9
   (B) 18
   (C) 27
   (D) 36

12. Which one of the following represents sufficient information to conclude that two triangles are congruent?

   (A) Two pairs of congruent sides
   (B) Two pairs of congruent angles and a pair of congruent sides
   (C) Three pairs of congruent angles
   (D) Two pairs of congruent sides and a pair of congruent angles

13. The area of a rectangle is 56. If the length is 16, what is the width?

   (A) 3.5
   (B) 7.5
   (C) 20
   (D) 40

GO ON

14. Which one of the following could represent the sides of an isosceles triangle?

   (A)   1, 7, and 8

   (B)   8, 5, and 5

   (C)   10, 12, and 14

   (D)   6, 6, and 15

15. Consider the following line segment $\overline{QT}$, on which points R and S lie.

   If $QS = 18$, $RT = 15$, and $QT = 23$, what is the length of $\overline{RS}$?

   (A)   13

   (B)   12

   (C)   11

   (D)   10

Answer Key—Arithmetic Diagnostic Test—Georgia CRCT

1. (C)	2. (B)	3. (B)	4. (A)
5. (A)	6. (D)	7. (B)	8. (A)
9. (C)	10. (B)	11. (D)	12. (B)
13. (A)	14. (C)	15. (B)	16. (D)
17. (C)	18. (C)	19. (D)	20. (D)
21. (C)	22. (B)	23. (D)	24. (C)
25. (A)			

Answer Key—Algebra Diagnostic Test—Georgia CRCT

1. (D)	2. (A)	3. (B)	4. (C)
5. (C)	6. (B)	7. (A)	8. (B)
9. (D)	10. (C)	11. (C)	12. (C)
13. (D)	14. (B)	15. (A)	16. (A)
17. (B)	18. (B)	19. (D)	20. (A)
21. (C)	22. (B)	23. (D)	24. (B)
25. (D)			

Answer Key—Geometry Diagnostic Test—Georgia CRCT

1. (B)   2. (D)   3. (C)   4. (C)

5. (A)   6. (B)   7. (C)   8. (A)

9. (B)   10. (D)   11. (D)   12. (B)

13. (A)   14. (B)   15. (D)

Chapter 1: Numbers and Operations

Welcome to Chapter One. In this chapter you will learn about the following topics:

a)   Squares and square roots

b)   Rational and irrational numbers

c)   Properties of exponents

d)   Scientific notation

As an example, if you knew the area of a square field was 225 square meters, how many feet of fencing would be needed to surround the field? We can use the properties of square roots to solve problems like this one and many others.

Later in the chapter, you will learn about exponents and their properties. Once you understand how to use exponents, you can apply them to solving sophisticated problems. Imagine calculating the weight of the earth – in ounces! It's easy when you know how to use exponents.

The product of two equal quantities is called a square. 81 is a square because $9 \times 9 = 81$. Another way to write 9×9 is 9^2. Fractions can be squares, too. $\dfrac{36}{49}$ is a square because $\dfrac{6}{7} \times \dfrac{6}{7} = \dfrac{36}{49}$. The opposite of a square is a square root. Since 81 is the square of 9, we can say that 9 is the square root of 81. The symbol for finding the square root of a number is "$\sqrt{}$" and is called a radical. Thus, $\sqrt{81} = 9$.

It is helpful to remember a few squares and square roots:

$0^2 = 0$

$1^2 = 1$	$6^2 = 36$	$11^2 = 121$
$2^2 = 4$	$7^2 = 49$	$12^2 = 144$
$3^2 = 9$	$8^2 = 64$	$13^2 = 169$
$4^2 = 16$	$9^2 = 81$	$14^2 = 196$
$5^2 = 25$	$10^2 = 100$	$15^2 = 225$

$\sqrt{0} = 0$	$\sqrt{36} = 6$	$\sqrt{144} = 12$
$\sqrt{1} = 1$	$\sqrt{49} = 7$	$\sqrt{169} = 13$
$\sqrt{4} = 2$	$\sqrt{64} = 8$	$\sqrt{196} = 14$
$\sqrt{9} = 3$	$\sqrt{81} = 9$	$\sqrt{225} = 15$
$\sqrt{16} = 4$	$\sqrt{100} = 10$	
$\sqrt{25} = 5$	$\sqrt{121} = 11$	

Once you have mastered a few squares and square roots, certain patterns become evident.

Try it!

Example 1:

$\sqrt{900} = ?$

Solution:

$\sqrt{900} = 30$. Since $3 \times 3 = 9$, then $30 \times 30 = 900$.

Let's consider the problem about the square field that was given in the beginning of this chapter.

Example 2:

A square lot has an area of 225 square meters.

Figure 1.1

How many meters of fencing are needed to enclose the space?

Solution:

The area of a square is found by multiplying two of its sides. Since all of the sides are equal, take the square root of the area to find the length of one side, which is $\sqrt{225} = 15$. Then the perimeter equals four times the length of one side. Multiply 15 by 4 to get the perimeter. $15 \times 4 = 60$. The square lot requires 60 meters of fencing to surround the property.

Numbers such as 1, -2, 0, and -3.7 all have positions on a number line:

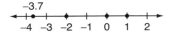

Figure 1.2

Square Roots and Number Lines

Square roots can also be shown on a number line.

Try it!

Example 3:

Place each number in its correct place on a number line:
$$\sqrt{4}, -2\frac{1}{2}, -\frac{2}{3}, \sqrt{49}, \sqrt{0}$$

Solution:

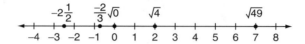

Figure 1.3

Notice that $\sqrt{0} = 0$ because $0 \times 0 = 0$.

Positive and Negative Square Roots

We showed earlier that $\sqrt{49} = 7$ because $7 \times 7 = 49$. However, it is true that $(-7) \times (-7) = 49$. Every positive number has two square roots, one positive and one negative. To avoid confusion, the negative square root is shown by placing a negative sign to the left of the radical. Thus, $\sqrt{49} = 7$, but $-\sqrt{49} = -7$.

It is important to remember that the negative sign cannot be under the radical sign, since $\sqrt{-49}$ has no solution in the real number system. The product of two positives or two negatives is always positive.

Try it!

Example 4:

Simplify the following expression.
$$\sqrt{81} + \left(-\sqrt{16}\right) + \sqrt{\frac{49}{64}}.$$

Solution:

$$9+(-4)+\frac{7}{8}=5\frac{7}{8}.$$

Rational and Irrational Numbers

Rational numbers can be expressed as the ratio of two integers. The following numbers are rational: $\frac{2}{3}$, 0.8, and 6. $\frac{2}{3}$ is rational because it is the ratio of 2 to 3.

0.8 is rational because it can expressed as $\frac{8}{10}$, or $\frac{4}{5}$ when it is reduced to simplest form.

6 is rational because it can be expressed as $\frac{6}{1}$.

Some rational numbers, when expressed as decimals, do not terminate. Here are two examples.

0.3333… or $0.\overline{3}$ and 0.142857142857… or $0.\overline{142857}$.

$0.\overline{3}$ is rational because it can be expressed as $\frac{1}{3}$.

$0.\overline{142857}$ is rational because it can be expressed as $\frac{1}{7}$.

The bar over the numbers means that those numbers repeat indefinitely.

Thus, rational numbers either terminate, such as 3.8, or repeat a pattern, such as 0.277277277…

A number that neither terminates nor repeats a pattern is called an irrational number. The following are examples of irrational numbers:

2.73841…, 0.86215…, 6.568319…

A well-known irrational number is the value of π (pi), which is approximately 3.141592…

The square roots of many numbers are irrational. Here are three examples:

$$\sqrt{17}=4.1213105\ldots,\quad \sqrt{82}=9.05538\ldots,\quad \text{and}\quad \sqrt{229}=15.1327\ldots$$

If the number under the radical is not a perfect square, then the answer will always be irrational.

Try it!

Example 5:

Indicate which numbers are rational and which are irrational.

$\sqrt{64}$ $\sqrt{7}$ $0.135135...$ $\sqrt{\dfrac{18}{50}}$

Solution:

$\sqrt{64}$ is rational because its value is $8\left(\text{or }\dfrac{8}{1}\right)$.

$\sqrt{7}$ is irrational because 7 is not a perfect square.

$0.135135...$ is rational because the decimal pattern repeats.

$\sqrt{\dfrac{18}{50}}$ is rational because it reduces to $\sqrt{\dfrac{9}{25}}$, which equals $\dfrac{3}{5}$.

Estimating Square Roots

Although $\sqrt{7}$ is irrational, it nonetheless has a place on the number line. Using a calculator, we can estimate the value of $\sqrt{7}$ to be about 2.65.

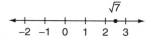

Figure 1.4

$\sqrt{7}$ lies between 2 and 3 and is slightly closer to 3.

Estimating square roots with irrational solutions requires patience and intelligence. Use the steps in Example 6 to estimate square roots.

Try it!

Example 6:

To the nearest tenth, calculate $\sqrt{10}$.

Solution:

Step 1: Decide which integers are located on each side of $\sqrt{10}$ by considering perfect squares.

$\sqrt{9} = 3, \sqrt{10} = ?, \sqrt{16} = 4$. So, $\sqrt{10}$ is located between 3 and 4.

Step 2: Decide which integer is closer to $\sqrt{10}$. Since 10 is much closer to 9 than it is to 16, $\sqrt{10}$ must be closer to 3 than 4.

Step 3: Estimate the value of $\sqrt{10}$ to the nearest tenth. Consider that $3.1 \times 3.1 = 9.61$ and $3.2 \times 3.2 = 10.24$. Since 10.24 is closer to 10 than is 9.61, $\sqrt{10}$, rounded to the nearest tenth, is 3.2.

Simplifying Square Roots

Some square roots, though irrational, can still be simplified.

Try it!

Example 7:

Simplify $\sqrt{18}$

Solution:

$\sqrt{18} = \sqrt{9} \times \sqrt{2} = 3\sqrt{2}$

Simplifying some square root problems can be more challenging.

Example 8:

Simplify $\sqrt{147}$

Solution:

Use a factor tree to identify perfect squares, if they exist.

147

49×3

$\sqrt{147} = \sqrt{49} \times \sqrt{3} = 7\sqrt{3}$

Adding and Subtracting Square Roots

When adding or subtracting square roots with the same number under the square root sign, we simply combine the numbers outside the square root sign.

Try it!

Example 9:

Simplify $6\sqrt{7} - 5\sqrt{7} + \sqrt{7}$.

Solution:

$6\sqrt{7} - 5\sqrt{7} + 1\sqrt{7} = 2\sqrt{7}$

It is important to note that $\sqrt{7}$ is the same as $1\sqrt{7}$, even though the 1 rarely appears.

Square roots can be added or subtracted if the number under the radical sign is the same.

Multiplying and Dividing Square Roots

When multiplying square roots, simply multiply the numbers under the radicals and simplify as needed.

Try it!

Example 10:

$$\sqrt{8} \times \sqrt{18} = ?$$

Solution:

$$\sqrt{8} \times \sqrt{18} = \sqrt{144} = 12$$

Example 11:

$$\sqrt{8} \times \sqrt{6} = ?$$

Solution:

$$\sqrt{8} \times \sqrt{6} = \sqrt{48} = \sqrt{16} \times \sqrt{3} = 4\sqrt{3}$$

When dividing square roots, divide the numbers under the radicals.

Example 12:

$$\sqrt{192} \div \sqrt{3} = ?$$

Solution:

$$\sqrt{192} \div \sqrt{3} = \sqrt{64} = 8$$

We conclude this chapter with a discussion of exponents. An exponent (also known as a power) is the number of times the base is used as a factor. In the expression 8^2, 8 is the base and 2 is the exponent. Exponents are useful tools when expressing certain quantities.

Example 13:

$32 = 2^?$

Solution:

$32 = 2 \times 2 \times 2 \times 2 \times 2 = 2^5$

Properties of Exponents

Consider the multiplication problem $8 \times 4 = 32$. Since $2^3 = 8$, $2^2 = 4$, and $2^5 = 32$, we can write this problem as $2^3 \times 2^2 = 2^5$. Notice that when multiplying numbers with the same bases, we add the exponents. Similarly, consider the division problem $32 \div 4 = 8$. We can write this problem as $2^5 \div 2^2 = 2^3$. Thus, when dividing numbers with the same base, we subtract the exponents.

Let's see what happens when we raise a power to another power.

Try it!

Example 14:

What is the value of $(2^3)^2$?

Solution:

Calculate the quantity in the parentheses : $2^3 = 8$. Raise 8 to the second power: $8^2 = 64$. Now 64 can be expressed as : $2 \times 2 \times 2 \times 2 \times 2 \times 2 = 2^6$. Therefore: $(2^3)^2 = 2^{3\times2} = 2^6 = 64$.

When raising one power to another, multiply the exponents.

Let's put all the properties of exponents into one problem.

Example 15:

Simplify the following: $\dfrac{(7^5)^2(7^3)}{7^{11}}$

Solution:

$(7^5)^2 = 7^{10}$. Then $\dfrac{7^{10} \times 7^3}{7^{11}} = \dfrac{7^{10+3}}{7^{11}} = \dfrac{7^{13}}{7^{11}} = 7^2 = 49$

Scientific Notation

A number uses scientific notation when it is expressed as $a \times 10^n$. The variable, a, must satisfy $1 \leq a < 10$ and n must be an integer.

Try it!

Example 16:

Express 2,014.7 in scientific notation.

Solution:

Step 1: Move the decimal point three places to the left so it is greater than or equal to 1 but less than 10:
2,014.7 → 2.0147

Step 2: Since we moved the decimal point three places to the left, we must multiply 2.0147 by 1000, which is 10^3.
Thus, 2,014.7, expressed in scientific notation, is 2.0147×10^3.

Example 17:

The distance from Earth to the Andromeda galaxy is approximately 2,200,000 light years. Each light year is about 6,000,000,000,000 miles. Express the distance to the Andromeda galaxy, in miles, in scientific notation.

Solution:

Convert both numbers into scientific notation and multiply:
$2{,}200{,}000 = 2.2 \times 10^6$ Likewise, $6{,}000{,}000{,}000{,}000 = 6 \times 10^{12}$
$(2.2 \times 10^6)(6 \times 10^{12}) = 13.2 \times 10^{18}$.

This answer is not in scientific notation. Since 13.2 is not less than 10, move the decimal point one place to the left and add another power of 10: The correct answer is 1.32×10^{19}.

Scientific notation can also be used to express very small numbers.

Example 18:

> The width of an atom is 0.00000002 centimeters. Express the width in scientific notation.

Solution:

Move the decimal point eight places to the right so that the first number is less than 10 but greater than or equal to 1: $0.00000002 \rightarrow 2.0$.

Since we moved the decimal point 8 places to the right, multiply 2.0 by 10^{-8}. Note that $10^{-8} = 0.00000001$. Thus, the width of an atom, expressed in scientific notation, is 2.0×10^{-8}.

Using Technology

Scientific and graphing calculators can be very useful when approximating square roots and converting numbers into scientific notation.

Try it!

Example 19:

> Use a calculator to express $\sqrt{21}$ to the nearest thousandth.

Solution:

Pressing the button with the radical and the number 21, the calculator shows 4.58257569496. Rounded to the nearest thousandth, the answer is 4.583.

Example 20:

> Multiply 3,234,111 and 17,636,087 and express the answer in scientific notation.

Solution:

After multiplying the two numbers, the calculator shows

5.70370629637 E 13.

On a calculator, the letter "E" and the number immediately following it represent the power of 10 that is required to express the number in scientific notation. Thus, $3,234,111 \times 17,636,087 = 5.70370629637 \times 10^{13}$.

Quiz for Chapter 1

1. $\sqrt{2500} = ?$

   (A) 500
   (B) 50
   (C) 5
   (D) 250

2. What is the perimeter of a square that has an area of 121 square inches?

   (A) 11
   (B) 22
   (C) 44
   (D) 62.5

3. What is the correct order, from lowest to highest, of the following five numbers?
   $\sqrt{16}, 7, -2.3, 1\frac{1}{2}, -\sqrt{10}$

   (A) $-2.3, -\sqrt{10}, 1\frac{1}{2}, 7, \sqrt{16}$

   (B) $-2.3, -\sqrt{10}, 1\frac{1}{2}, \sqrt{16}, 7$

   (C) $-\sqrt{10}, -2.3, 1\frac{1}{2}, 7, \sqrt{16}$

   (D) $-\sqrt{10}, -2.3, 1\frac{1}{2}, \sqrt{16}, 7$

4. What is the value of $\sqrt{36} + \left(-\sqrt{9}\right) + \sqrt{\frac{1}{169}}$?

   (A) $3\frac{1}{13}$

   (B) $2\frac{12}{13}$

   (C) $27\frac{1}{169}$

   (D) $45\frac{1}{169}$

5. Which number is irrational?

    (A)   17.636363….

    (B) $\sqrt{196}$

    (C) $\sqrt{96}$

    (D) $\dfrac{313}{17}$

6. Between which two integers is the value of $\sqrt{29}$?

    (A)   3 and 4
    (B)   4 and 5
    (C)   5 and 6
    (D)   6 and 7

7. What is the simplified expression for $\sqrt{75}$?

    (A) $3\sqrt{5}$
    (B) $5\sqrt{3}$
    (C)   37.5
    (D) $\dfrac{25}{3}$

8. What is the simplified expression for $2\sqrt{3}+\sqrt{27}$?

    (A) $2\sqrt{30}$
    (B)   16.5
    (C) $5\sqrt{3}$
    (D)   30

9. What is the simplified expression for $\sqrt{20}\times\sqrt{45}$?

    (A)   450
    (B)   30
    (C) $90\sqrt{10}$
    (D)   300

10. What is the simplified expression for $\sqrt{180}\div\sqrt{5}$?

    (A) $\sqrt{36}$
    (B)   30
    (C)   6
    (D)   16

11. What is the simplified expression for $\dfrac{(5^2)^3(5^3)}{5^8}$?

    (A) 50

    (B) 125

    (C) 5

    (D) 0.5

12. What is the number 3,745.6 written in scientific notation?

    (A) 3.7456×10^3

    (B) $37,456 \times 10^1$

    (C) 37.456×10^2

    (D) 3745.6×10^0

13. What is the value of $\dfrac{(20)(315)}{0.0045}$, expressed in scientific notation?

    (A) 1.4×10^5

    (B) 1.4×10^6

    (C) 14×10^5

    (D) 14×10^6

14. The perimeter of a square is 48 feet. What is its area in square feet?

    (A) 144

    (B) 24

    (C) $\sqrt{48}$

    (D) $\sqrt{12}$

Answer Key

1. **(B)** $50 \times 50 = 2500$. Therefore, $\sqrt{2500} = 50$.

2. **(C)** $\sqrt{121} = 11$, which is the length of one side of the square. A square has four equal, or congruent, sides, so multiply 11 by 4 to get 44.

3. **(D)** Since $-\sqrt{10} \approx -3.16$ and $\sqrt{16} = 4$, the correct order is $-\sqrt{10}$, -2.3, $1\frac{1}{2}$, $\sqrt{16}$, 7.

4. **(A)** $\sqrt{36} + (-\sqrt{9}) + \sqrt{\frac{1}{169}} = 6 - 3 + \frac{1}{13} = 3\frac{1}{13}$.

5. **(C)** Since 96 is not a perfect square, $\sqrt{96}$ is irrational.

6. **(C)** Since $\sqrt{25} = 5$ and $\sqrt{36} = 6$, $\sqrt{29}$ lies between 5 and 6.

7. **(B)** $\sqrt{75} = \sqrt{25} \times \sqrt{3} = 5\sqrt{3}$.

8. **(C)** $2\sqrt{3} + \sqrt{27} = 2\sqrt{3} + 3\sqrt{3} = 5\sqrt{3}$.

9. **(B)** $\sqrt{20} \times \sqrt{45} = \sqrt{900} = 30$.

10. **(C)** $\sqrt{180} \div \sqrt{5} = \sqrt{36} = 6$.

11. **(C)** $\dfrac{(5^2)^3(5^3)}{5^8} = \dfrac{(5^6)(5^3)}{5^8} = \dfrac{5^9}{5^8} = 5^1 = 5$.

12. **(A)** Move the decimal point to the left: $3{,}745.6 = 3.7456 \times 10^3$.

13. **(D)** First express each number in scientific notation: $20 = 2 \times 10^1$, $315 = 3.15 \times 10^2$, and $0.0045 = 4.5 \times 10^{-3}$. Then $\dfrac{(20)(315)}{0.0045} = \dfrac{(2 \times 10^1)(3.15 \times 10^2)}{4.5 \times 10^{-3}} = \dfrac{6.3 \times 10^3}{4.5 \times 10^{-3}}$ $= 1.4 \times 10^6$.

14. **(A)** Each side of the square is 12 feet. Then $(12)(12) = 144$ square feet.

Chapter 2: Geometry

Welcome to Chapter Two. In this chapter you will learn about the following topics:

a)   Parallel and perpendicular lines

b)   Angles formed by parallel lines

c)   The Pythagorean theorem

Your goal is to be able to understand these geometric concepts and apply them to algebra problems.

Before discussing parallel lines, a basic geometry review is in order. We will start by discussing points, lines and planes.

Points

Point: A **point** is the simplest geometric concept. It has no dimension and is usually described by a letter, such as point P.

Lines

Line: A **line** is a figure that extends infinitely in opposite directions.

Figure 2.1

The line above can be described as $\overleftrightarrow{AB}$ or $\overleftrightarrow{BA}$. Sometimes a line is described with a single, lowercase letter.

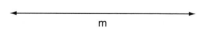

Figure 2.2

Two points are required to determine a line. A line has one dimension (length) and contains an infinite number of points.

Three points that lie on the same line are called collinear. Points A, B and C on the line below are collinear.

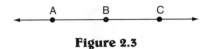

Figure 2.3

Non-collinear points are points that do not lie on the same line.

The points X, Y and Z below are non-collinear points.

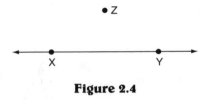

Figure 2.4

Planes

Plane: A **plane** is a two-dimensional figure that extends infinitely in two directions.

It may be useful to think of a white board or a piece of paper as a plane if you can imagine each one extending to infinity. Three non-collinear points determine a plane. In Figure 2.5, we have plane ABC.

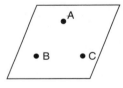

Figure 2.5

A plane can be also be described by one uppercase letter. Figure 2.6 illustrates plane M.

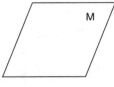

Figure 2.6

Points and lines that lie in the same plane are called **coplanar**. Our discussion of parallel and perpendicular lines assumes all lines are coplanar.

Parallel Lines

Parallel lines are lines that do not intersect. Railroad tracks and the stripes on the American flag remind us of parallel lines. Lines that are parallel are denoted by the symbol "$\parallel$".

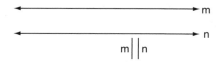

Figure 2.7

"$m \parallel n$" means line m is parallel to line n.

A line that intersects two parallel lines is called **a transversal**. In the figure below, line l is a transversal.

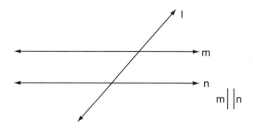

Figure 2.8

When a transversal intersects parallel lines, it creates pairs of angles that have equal measures or are supplementary (add up to 180°).

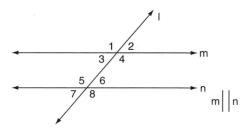

Figure 2.9

$\angle 1$ and $\angle 5$ have equal measures and are called **corresponding angles**. This is also true for the following three pairs of angles: (a) $\angle 2$ and $\angle 6$, (b) $\angle 3$ and $\angle 7$, and (c) $\angle 4$ and $\angle 8$.

$\angle 3$ and $\angle 6$ have equal measures and are called **alternate** (or opposite) **interior angles**. This is also true for $\angle 4$ and $\angle 5$.

$\angle 2$ and $\angle 7$ have equal measures and are called **alternate** (or opposite) **exterior angles**. This is also true for $\angle 1$ and $\angle 8$.

$\angle 1$ and $\angle 4$ have equal measures and are called **vertical angles**.

This is also true for the following three pairs of angles: (a) $\angle 2$ and $\angle 3$, (b) $\angle 5$ and $\angle 8$, and (c) $\angle 6$ and $\angle 7$.

All other pairs of angles are **supplementary**. Many math textbooks emphasize that $\angle 3$ and $\angle 5$ are supplementary and are designated as "consecutive interior angles."

Refer to Figure 2.9 for the following two examples.

Try it!

Example 1:

$m \angle 2$ (read "the measure of $\angle 2$") $= 80°$, $m \angle 7 = (5x)°$. What is the value of x?

Solution:

$\angle 2$ and $\angle 7$ have equal measures because they are alternate exterior angles. Then $5x = 80$, so $x = \dfrac{80}{5} = 16$.

NOTE: The symbol "m" can be used to name a line. However, "m" is also the abbreviation for "measure" when naming an angle. The meaning of "m" can be understood from the context in which it is given.

Example 2:

$m \angle 3 = (2x)°$ and $m \angle 8 = (3x)°$. What is the value of x?

Solution:

$\angle 3$ and $\angle 8$ are supplementary angles, so set their sum equal to $180°$. Then $(3x) + (2x) = 180$. So, $5x = 180$. Thus, $x = \dfrac{180}{5} = 36$.

Perpendicular Lines

Lines that are **perpendicular** intersect to form a right angle, which is $90°$. The symbol "$\perp$" means perpendicular.

In the figure below, $\overleftrightarrow{AB} \perp \overleftrightarrow{CD}$

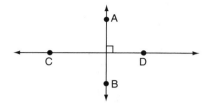

Figure 2.10

Try it!

Example 3:

In the diagram below, $l \parallel m$ and $t \perp l$

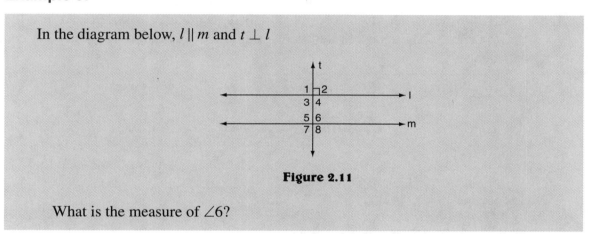

Figure 2.11

What is the measure of $\angle 6$?

Solution:

Since $l \parallel m$, $m \angle 2 = m \angle 6$ since they are corresponding angles. Then $m \angle 2 = 90°$, since $t \perp l$. Therefore the $m \angle 6$ must also equal 90°.

We have just proven another theorem in Geometry. Because $m \angle 6 = 90°$, $t \perp m$. Thus, if a line is perpendicular to one of two parallel lines, it is also perpendicular to the other line.

In this figure, $l \parallel m$, $t \perp l$, and $t \perp m$.

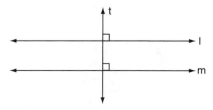

Figure 2.12

Congruent Figures

Observe the two triangles below.

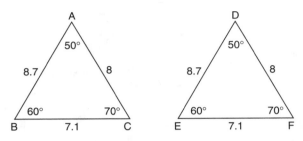

Figure 2.13

$\triangle ABC$ and $\triangle DEF$ are called **congruent triangles**. Essentially, one triangle may be placed directly on top of the other triangle so that they appear identical. Each side and angle of $\triangle ABC$ is congruent to a corresponding side and angle in $\triangle DEF$. We use the symbol "$\cong$" to denote congruency. In the triangles above, we can make the following congruency statements:

$\triangle ABC \cong \triangle DEF$ $\qquad$ $\angle A \cong \angle D$ $\qquad$ $\angle B \cong \angle E$ $\qquad$ $\angle C \cong \angle F$

$\overline{AB} \cong \overline{DE}$ $\qquad\qquad$ $\overline{BC} \cong \overline{EF}$ $\qquad\qquad$ $\overline{AC} \cong \overline{DF}$

Try it!

Example 4:

In the figure below, $\triangle FGH \cong \triangle JKL$

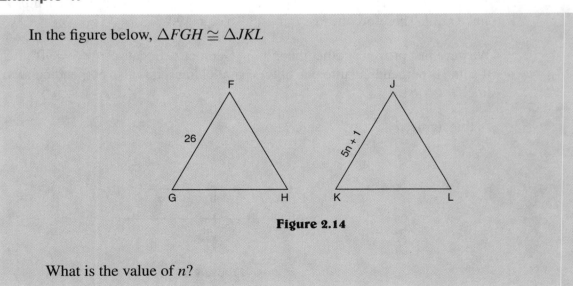

Figure 2.14

What is the value of n?

Solution:

Since $\triangle FGH \cong \triangle JKL$, then $\overline{FG} \cong \overline{JK}$ So, $26 = 5n + 1$. By subtracting 1 from each side, this equation simplifies to $25 = 5n$. Thus, $n = \dfrac{25}{5} = 5$.

Right Triangles

A right triangle contains a right angle. The longest side of a right triangle is called the hypotenuse and the other two sides are called legs. The hypotenuse is opposite the right angle.

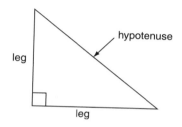

Figure 2.15

Try it!

Example 5:

In the right triangle below, $m \angle A = m \angle C$. What is the measure of each angle?

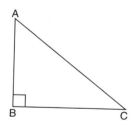

Figure 2.16

Solution:

The sum of the measures of the angles in a triangle is 180°. Since $m \angle B = 90°$, let x equal the measure of each of the congruent angles. Then $90 + x + x = 180$. This equation simplifies to $90 + 2x = 180$. Then $2x = 180 - 90 = 90$. Thus, $x = \dfrac{90}{2} = 45$. This means that each of $\angle A$ and $\angle C$ has a measure of 45°.

The Pythagorean Theorem

The **Pythagorean theorem** states that the sum of the squares of the legs equals the square of the hypotenuse.

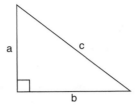

Figure 2.17

With respect to this figure, $a^2 + b^2 = c^2$.

Try it!

Example 6:

Consider the diagram shown below.

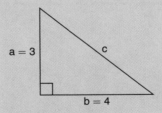

Figure 2.18

What is the measure of c?

Solution:

$3^2 + 4^2 = c^2$. Then $9 + 16 = c^2$, which simplifies to $25 = c^2$. Thus, $c = \sqrt{25} = 5$.

Notice that all the sides are integers. Right triangles with sides that are integers are called **Pythagorean triples**. Some common Pythagorean triples are 3-4-5, 5-12-13, and 8-15-17.

Furthermore, multiples of triples are also triples.

Example 7:

Consider the diagram shown below.

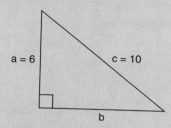

Figure 2.19

What is the measure of b?

Solution:

$6^2 + b^2 = 10^2$. Then $36 + b^2 = 100$ which simplifies to $b^2 = 64$.

Thus, $b = \sqrt{64} = 8$.

Since 3-4-5 is a triple, so is 6-8-10.

In Chapter One, we discovered that squares of numbers can be pictured as geometric squares.

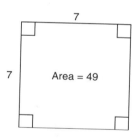

Figure 2.20

We can use this fact to demonstrate the Pythagorean theorem graphically, namely $a^2 + b^2 = c^2$.

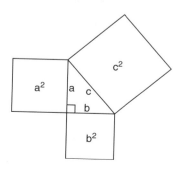

Figure 2.21

Quiz for Chapter 2

1. A point has how many dimensions?

   (A) 0
   (B) 1
   (C) 2
   (D) 3

2. Which statement is true concerning perpendicular lines?

   (A) They never intersect.
   (B) They intersect to form an angle less than 90°.
   (C) They intersect to form an angle greater than 90°.
   (D) They intersect to form right angles.

3. Three points that lie on the same line are called _____ points.

   (A) non-collinear
   (B) non-coplanar
   (C) collinear
   (D) intersecting

4. Consider the following figure.

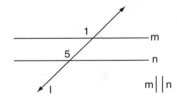

Figure 2.22

   ∠1 and ∠5 are called _____ angles.

   (A) alternate exterior
   (B) corresponding
   (C) alternate interior
   (D) vertical

5. Consider the following figure.

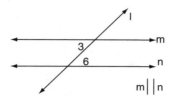

Figure 2.23

∠3 and ∠6 are called _____ angles.

(A)   alternate interior
(B)   inverse angles
(C)   supplementary
(D)   complementary

6. Consider the following figure.

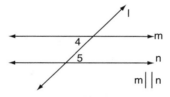

Figure 2.24

If $m\angle 5 = 68°$ and $m\angle 4 = (4x)°$, what is the value of x?

(A)   180
(B)   68
(C)   34
(D)   17

7. Triangles whose corresponding sides and angles have equal measures are called
_____.

(A)   congruent
(B)   supplementary
(C)   complementary
(D)   corresponding

8. Consider the following figures.

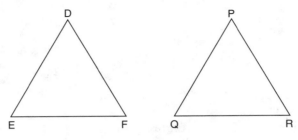

Figure 2.25

$\triangle DEF \cong \triangle PQR$. If $m\angle D = 4x - 20$, and $m\angle P = 2x + 30$, what is the measure of $\angle P$?

(A) 25°
(B) 40°
(C) 70°
(D) 80°

9. Consider right $\triangle WXY$, as shown below.

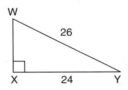

Figure 2.26

What is the measure of $\overline{WX}$?

(A) 5
(B) 10
(C) 12
(D) 20

10. Consider $\triangle ABC$, as shown below.

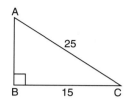

Figure 2.27

What is the perimeter of this triangle?

(A)   20
(B)   40
(C)   60
(D)   80

Answer Key

1. (A)   A point has no (or zero) dimension.

2. (D)   Perpendicular lines intersect to form right angles.

3. (C)   Three points that lie on the same line are called collinear.

4. (B) $\angle 1$ and $\angle 5$ are corresponding angles, since they lie in the same position when a transversal intersects the two lines containing them.

5. (A) $\angle 3$ and $\angle 6$ are alternate interior angles, since they lie on opposite sides of the transversal and between the parallel lines.

6. (D) $\angle 5$ and $\angle 4$ have equal measures because they are alternate interior angles, so $4x = 68$. Thus, $x = \dfrac{68}{4} = 17$.

7. (A)   Triangles with corresponding sides of equal length and angles of equal measures are called congruent triangles.

8. (D)   Since $\triangle DEF \cong \triangle PQR$, $m\angle D = m\angle P$. Then $4x - 20 = 2x + 30$. After subtracting $2x$ from both sides, we get $2x - 20 = 30$. So, $2x = 50$. Thus, $x = \dfrac{50}{2} = 25$. Then $m\angle P = 2x + 30 = (2)(25) + 30 = 80°$.

9. (B)   Let $WX = a$. By the Pythagorean theorem, $a^2 + 24^2 = 26^2$. Then $a^2 + 576 = 676$, which simplifies to $a^2 = 100$. Thus, $a = \sqrt{100} = 10$. (Note that $10 - 24 - 26$ is a multiple of the $5 - 12 - 13$ Pythagorean triple.)

10. (C)   By the Pythagorean theorem, $a^2 + 20^2 = 25^2$, which becomes $a^2 + 400 = 625$. Then $a^2 = 225$, so $a = \sqrt{225} = 15$. Thus, the perimeter is $15 + 20 + 25 = 60$. (Note that $15 - 20 - 25$ is a multiple of the $3 - 4 - 5$ Pythagorean triple.)

Chapter 3: Algebra—Part 1

Welcome to Chapter Three. In this chapter you will learn about the following topics:

a)   Algebraic expressions

b)   Equations in one variable

c)   Absolute value

d)   Inequalities

In the next few chapters, we will learn to use algebra skills to solve word problems and equations. Your goals in this chapter are to use algebraic expressions to model real-life situations, solve equations that use one variable, and understand absolute value and inequalities.

The skills you learn in this chapter will be used for the rest of your academic career.

Algebraic Expressions

Suppose that Brittany has $8.00 more than Joshua. How much money does Brittany have? The answer is that you cannot know how much money Brittany has until you know how much Joshua has. If Joshua has $12.00, then Brittany has $20.00 because $12.00 + $8.00 = $20.00. We can express the money Brittany has by using a variable.

A **variable** is a letter or symbol that represents an unknown quantity. Letters such as x or n are frequently used, although any letter can be a variable. Mathematical expressions that use one or more variables are called **algebraic expressions**. Let's say that n represents the money Joshua has. We can now create an algebraic expression that represents the amount of money Brittany has.

If $n =$ the amount of money Joshua has, then $n + 8 =$ the amount of money Brittany has. We add 8 to n because Brittany has $8.00 more than Joshua. The expression "more than" means addition.

There are several other words and expressions that represent mathematical operations, as shown in the following list.

Expression	Operation
increased by	addition
added to	addition
plus	addition
decreased by	subtraction
minus	subtraction
less (or less than)	subtraction
times	multiplication
double, triple, etc.	multiplication
shared	division
split equally	division
cut	division

Certain words are the results of operations in mathematics. They are:

Sum – the answer to an addition problem

Difference – the answer to a subtraction problem

Product – the answer to a multiplication problem

Quotient – the answer to a division problem

Try it!

Example 1:

Use an algebraic expression to represent five less than m.

Solution:

"Less than" means subtraction. The answer is $m - 5$.

Example 2:

Use an algebraic expression to represent the product of n and 7.

Solution:

Product means multiply. The answer is $7 \times n$ or $7n$.

Example 3:

William shared n dollars equally with his three friends. How much did each person receive?

Solution:

Shared equally means division. The answer is $n \div 4$ or $\dfrac{n}{4}$. Note that we divided by 4, since William also gets a share of the n dollars.

Example 4:

Janice has four times as many downloaded songs as she had last year. If she had m songs last year, how many does she have this year?

Solution:

Four times as many means multiplication. The answer is $4 \times m$ or $4m$.

One-Step Equations

If two girls, each weighing 100 pounds, are seated on opposite ends of a seesaw, the seesaw will balance. What would happen if a 50-pound. girl sat in front of one of the girls? What would happen if another 50-pound girl sat on the other side? The image of a balancing seesaw will help us understand how to work with equations.

An **equation** is a mathematical statement that indicates two quantities are equal.

The following are examples of equations:

$x + 9 = 12$

$3x + 2 = 5x - 7$

$r \times t = d$

Try it!

Example 5:

$x + 5 = 8$. What is the value of x?

Solution:

It is obvious that x must equal 3 because $3 + 5 = 8$. However, in some equations, the solutions may not be so obvious.

Example 6:

$x + 19 = -371$. What is the value of x?

Most students cannot perform such sophisticated calculations mentally. However, if you use Algebra's two basic rules, you can solve any equation.

Rule 1: To isolate a variable, perform the opposite operation.

Rule 2: Whatever changes are made to one side of the equation must be done to the other side as well.

Let's go back to Example 6.

Solution:

Since $x + 19 = -371$, subtract 19 from both sides to isolate x. Then

$x + 19 - 19 = -371 - 19$. Finally, simplify both sides so that $x = -371 - 19 = -390$.

(Remember: $+19 - 19$ always equals 0.)

Let's solve some equations using subtraction, multiplication, and division.

Example 7:

What is the value of x in the equation $x - 17 = 43$?

Solution:

Solve for x by adding 17 to both sides. Then $x - 17 + 17 = 43 + 17$

Thus, $x = 60$.

Example 8:

What is the value of x in the equation $7x = 98$?

Solution:

Divide both sides by 7 to isolate the variable. Then $\dfrac{7x}{7} = \dfrac{98}{7}$, so $x = 14$.

Example 9:

What is the value of x in the equation $\dfrac{x}{15} = -12$?

Solution:

Multiply both sides by 15 to isolate the variable. Then $(15)(\dfrac{x}{15}) = (-12)(15)$. Thus, $x = -180$.

Multi-Step Equations

Solving equations often requires more than one step.

Try it!

Example 10:

What is the value of n in the equation $2n - 12 = -38$?

Solution:

Although both subtraction and multiplication are in this problem, we still use the two basic rules of Algebra to solve for n. To isolate n, add 12 to both sides of the equation to get $2n - 12 + 12 = -38 + 12$. Now simplify to get $2n = -26$. Finally, divide both sides by 2. Then $\dfrac{2n}{2} = \dfrac{-26}{2}$, which simplifies to $n = -13$.

You can check your solution by replacing it for the variable.

$2(-13) - 12 = -38$

$-26 - 12 = -38$

$-38 = -38$

Equations with Decimals

An equation can contain decimals.

Try it!

Example 11:

What is the value of n in the equation $-8.4 = 0.28n$?

Solution:

Divide both sides by 0.28 to isolate n. Then $\dfrac{-8.4}{0.28} = \dfrac{0.28n}{0.28}$.

When dividing decimals, move the decimal point in the divisor as many places to the right as needed to make it an integer. Remember to move the decimal point the same amount of places to the right in the number being divided. This means that $0.28\overline{)-8.4}$ becomes $28\overline{)-840}$. The answer is $n = -30$.

Equations with Fractions

Some equations contain fractions.

Try it!

Example 12:

What is the value of r in the equation $\dfrac{2}{3}r = 10$?

Solution:

Solve for r by multiplying both sides of the equation by $\dfrac{3}{2}$, the reciprocal of $\dfrac{2}{3}$. Then $\dfrac{3}{2}(\dfrac{2}{3}r) = \dfrac{3}{2}(10)$. Thus, $r = 15$.

Multi-Variable Equations

Equations can be solved for a particular variable, even if they have more than one variable.

Try it!

Example 13:

> The expression "Rate × Time = Distance" is often abbreviated as $r \times t = d$. What is the expression for t in terms of r and d ?

Solution:

Divide both sides by r. Then $\dfrac{r \times t}{r} = \dfrac{d}{r}$. Thus, $t = \dfrac{d}{r}$.

Absolute Value

The **absolute value** of a number is the number of units a quantity is from 0. The absolute value uses the symbol "| |". When you see $|x|$, it is read "the absolute value of x."

The absolute value of any nonzero number is always expressed as a positive number. For example, $|3| = 3$ and $|-3| = 3$. The absolute value of zero is zero.

Try it!

Example 14:

> What is the value of $|-7| + 3$?

Solution:

$|-7| + 3 = 7 + 3 = 10$.

When finding the absolute value of a variable, there will be two answers (unless the variable equals zero.)

Example 15:

What are the values of x in the equation $|x| - 7 = 12$?

Solution:

Add 7 to both sides of the equation in order to isolate the variable. Then $|x| - 7 + 7 = 12 + 7$. This simplifies to $|x| = 19$. Because $|x|$ equals 19, the answers are $x = 19$ or $x = -19$.

Inequalities

An **inequality** states that one quantity is greater (or less) than another.

$m > n$ means "m is greater than n."

$-3 < -1$ means "-3 is less than -1."

$12r \geq -5.3$ means "$12r$ is greater than or equal to -5.3."

$-\dfrac{2}{3}x \leq 7$ means "$\dfrac{-2}{3}x$ is less than or equal to 7."

Inequalities share many of the same properties as equations.

Try it!

Example 16:

What is the solution for c in the inequality $4c + 12 \leq 16$?

Solution:

Subtract 12 from both sides of the inequality so that $4c + 12 - 12 \leq 16 - 12$. Now simplify to get $4c \leq 4$. Divide both sides of the inequality by 4, so that $\dfrac{4c}{4} \leq \dfrac{4}{4}$. Finally, the answer is $c \leq 1$.

It is important to note that when dividing or multiplying an inequality by a negative quantity, you must reverse the inequality sign.

Example 17:

What is the solution for p in the inequality $-\dfrac{2}{3}p > -12$?

Solution:

Multiply both sides of the equation by $-\dfrac{3}{2}$. Remember to reverse the inequality sign. Then $(-\dfrac{3}{2})(-\dfrac{2}{3}p < (-12)(-\dfrac{3}{2})$. Thus, $p < 18$.

Inequalities can be used to model real-life situations.

Example 18:

Julia needs at least \$20 to cover the cost of materials for a puppet show.

The cost is to be shared equally between Julia and the nine other students in the class. What is the least amount of money each student must spend?

Solution:

Create an inequality that models the situation.

Let n equal the least amount of money each student needs to spend. Since 10 students (Julia plus nine others) will share the cost, use the inequality $10n \geq 20$. Then $\dfrac{10}{10}n \geq \dfrac{20}{10}$, so $n \geq 2$.

We used the $\geq$ sign because the cost will be \$2 or more per student.

Inequalities can be graphed on number lines.

Example 19:

Graph the inequality $-3x > 12$ on a number line.

Solution:

First solve for x. Then $\dfrac{-3x}{-3} < \dfrac{12}{-3}$. (Remember to reverse the sign of the inequality because we have divided it by a negative number.) The solution is $x < -4$.

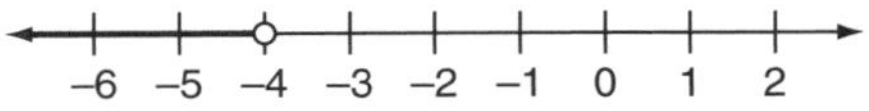

Figure 3.1

Notice the graph above. All of the numbers that are less than -4 have been darkened. Pay close attention to the open circle on the number -4. This indicates that all numbers less than -4, but not equal to -4, are part of the solution.

Example 20:

Graph the inequality $5x \geq 20$ on a number line.

Solution:

First solve for x. Then $\dfrac{5x}{5} \geq \dfrac{20}{5}$, so $x \geq 4$.

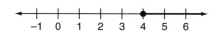

Figure 3.2

Observe that all the numbers greater than 4 have been darkened. Further, since the inequality indicates that x is greater than or equal to 4, the number 4 is part of the graph.

Thus, the circle at 4 is also darkened.

Quiz for Chapter 3

1. Which expression represents the phrase "the product of 5 and n decreased by 7"?

   (A) $5 + n - 7$

   (B) $\dfrac{5}{n} - 7$

   (C) $5n - 7$

   (D) $7 - 5n$

2. The result of a division problem is the _____.

   (A) quotient

   (B) product

   (C) sum

   (D) difference

3. What is the value of x in the equation $x - (-15) = -12$?

   (A) -3

   (B) -27

   (C) 3

   (D) 27

4. What is the value of x in the equation $-2x = -47$?

   (A) 49

   (B) 23.5

   (C) -23.5

   (D) -49

5. What is the value of x in the equation $\dfrac{2}{3}x + 5 = 9$?

   (A) -21

   (B) $\dfrac{8}{3}$

   (C) 6

   (D) 21

6. What is the value of r in the equation $-2.7r = 0.729$?

   (A) 2.7

   (B) 0.027

   (C) -0.27

   (D) -27

7. Given $C = 2\pi r$, then $r =$ _____

   (A) $\dfrac{2\pi}{C}$

   (B) $C - 2\pi$

   (C) $2\pi - C$

   (D) $\dfrac{C}{2\pi}$

8. What is (are) the value(s) of n in the equation $|n| - 12 = 14$?

   (A) Only 26
   (B) Only -26
   (C) -26 and 26
   (D) Only 2

9. What is the solution for x in the inequality $3x \geq 12$?

   (A) $x > 4$
   (B) $x < 4$
   (C) $x \leq 4$
   (D) $x \geq 4$

10. What is the solution for x in the inequality $-4x - 4 > 14$?

    (A) $x > -4.5$
    (B) $x < -4.5$
    (C) $x > 4.5$
    (D) $x < 4.5$

11. Look at the following graph on the number line.

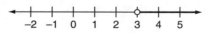

Figure 3.3

This represents the solution to which of the following inequalities?

    (A) $\dfrac{1}{3}x > 1$

    (B) $\dfrac{1}{3}x < 1$

    (C) $3x < 1$
    (D) $3x > 1$

12. Which of the following describes the graph on the number line of the inequality $-2n + 9 \geq 11$?

    (A)   An open circle on -1 and an arrow to the right.

    (B)   A closed circle on 1 and an arrow to the right.

    (C)   An open circle on 1 and an arrow to the left.

    (D)   A closed circle on -1 and an arrow to the left.

Answer Key

1. **(C)** A product is the result of multiplication and "decreased by" means "subtract." The answer is $5 \times n - 7$ or $5n - 7$.

2. **(A)** The quotient is the result of a division problem.

3. **(B)** $x - (-15) = -12$, $x + 15 = -12$, $x + 15 - 15 = -12 - 15$, thus $x = -27$.

4. **(B)** $-2x = -47$, $\dfrac{-2x}{-2} = \dfrac{-47}{-2}$, thus $x = 23.5$.

5. **(C)** $\dfrac{2}{3}x + 5 = 9$, $\dfrac{2}{3}x + 5 - 5 = 9 - 5$, $\dfrac{2}{3}x = 4$, $\left(\dfrac{3}{2}\right)\left(\dfrac{2}{3}x\right) = (4)\left(\dfrac{3}{2}\right)$, thus $x = 6$.

6. **(C)** $-2.7r = 0.729$, $\dfrac{-2.7r}{-2.7} = \dfrac{0.729}{-2.7}$, thus $r = -0.27$.

7. **(D)** $C = 2\pi r$, $\dfrac{C}{2\pi} = \dfrac{2\pi r}{2\pi}$, thus $\dfrac{C}{2\pi} = r$.

8. **(C)** $|n| - 12 = 14$, $|n| - 12 + 12 = 14 + 12$, $|n| = 26$, thus $n = -26$ or $n = 26$.

9. **(D)** $3x \geq 12$, $\dfrac{3x}{3} \geq \dfrac{12}{3}$, thus $x \geq 4$.

10. **(B)** $-4x - 4 > 14$, $-4x - 4 + 4 > 14 + 4$, $-4x > 18$, $\dfrac{-4x}{-4} < \dfrac{18}{-4}$, thus $x < -4.5$.

11. **(A)** The graph shows an open circle and an arrow to the right. Thus, its solution is $x > 3$. By dividing both sides of this inequality by 3, we get the equivalent inequality $\dfrac{1}{3}x > 1$.

12. **(D)** $-2n + 9 \geq 11$, $-2n + 9 - 9 \geq 11 - 9$, $-2n \geq 2$, $\dfrac{-2n}{-2} \leq \dfrac{2}{-2}$, thus $n \leq -1$.

The graph of this inequality is represented by a closed circle at -1 and an arrow to the left.

Chapter 4: Algebra—Part 2

Welcome to Chapter Four. In this chapter, you will learn about the following topics:

a) Relations

b) Functions

c) Forecasting data based on functions

Relations

A relation is any set of ordered pairs. The following is a relation.

$\{(2, 0), (-2, -5), (7, 4), (-3, 5)\}$

This same relation can be expressed graphically.

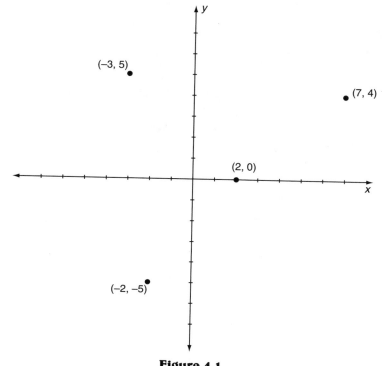

Figure 4.1

We can also express the relation in an input-output table.

Input	Output
2	0
−2	−5
7	4
−3	5

The set of all inputs is called the **domain**. The set of all outputs is called the **range**. The domain for the relation above is {−3, −2, 2, 7}. The range for the relation is {−5, 0, 4, 5}.

Functions

A **function** is a relation that establishes a relationship between inputs and outputs. It is important to note that for any input there is exactly one output.

Try it!

Example 1:

Does the following input-output table express a function?

Input	Output
6	−3
12	−6
18	−9
24	−12
30	−15

Solution:

The input-output table expresses a function because each input has exactly one output.

Example 2:

Does the following input-output table express a function?

Input	Output
7	2
3	8
5	−3
7	−7

Solution:

This table does not express a function. Notice that the input, 7, has two outputs, 2 and −7. Each input in a function can only have one output.

Example 3:

Does the following input-output table express a function?

Input	Output
7	3
8	3
9	3
23	3

Solution:

Although the outputs are identical, each input has only one output. Therefore, the input-output table expresses a function. Thus, in a function, a single output value may have more than one input value.

Let's see what a function looks like when it's graphed. We'll graph the points (−3, 3), (−2, 2), (−1, 1), (0, 0), (1, 1), (2, 2), and (3, 3).

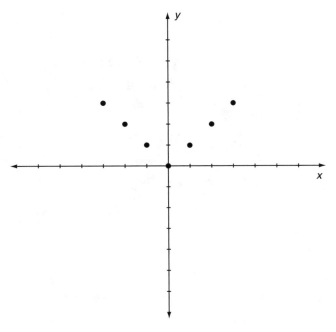

Figure 4.2

We can use a vertical line test to verify the graph is a function. If a vertical line passes through only one point on the graph, then the graph represents a function.

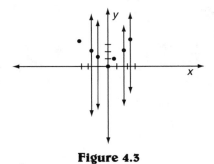

Figure 4.3

Example 4:

Which of the following graphs are functions?

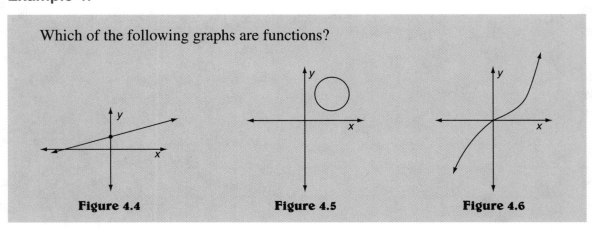

Figure 4.4   **Figure 4.5**   **Figure 4.6**

Solution:

Draw vertical lines through each graph.

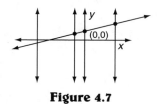

Figure 4.7

Thus, this graph is a function.

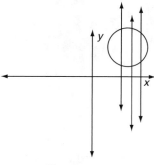

Figure 4.8

Thus, this graph is not a function.

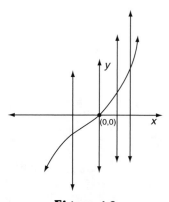

Figure 4.9

Thus, this graph is a function.

Functions are useful tools for drawing conclusions about data.

Example 5:

Brendan noticed that certain seats at a concert cost more than others. He created an input-output table showing the distance from the stage as the input and ticket cost as the output.

Input Distance in feet	Output Cost in $
80	45
120	35
150	27.50
200	25

Brendan wants to know:

a) Does the data represent a function?
b) What would a seat cost if it were 40 feet from the stage?
c) What conclusions can he draw from the data?

Solution:

a) Since each input has exactly one output, the data represent a function.

b) The closer a seat is to the stage, the more expensive the ticket cost will be.

Since a seat 40 feet from the stage is closer than a seat 80 feet from the stage, the ticket price should exceed $45.

c) Brendan can conclude that seats closer to the stage are more expensive because he assumes that the view and sound are better.

Functions can be used to predict outputs.

Example 6:

The following input-output table shows the number of shoelace holes in a shoe and the recommended length, in inches, of its shoelace.

Input Holes	Output Length
4	12
6	18
8	24

Based on this table, if a hiking boot has 18 holes, how long should its shoelace be?

Solution:

You may have noticed that the length of the shoelace is three times the number of holes. Therefore, the hiking boot will need shoelaces that are 54 inches long because $18 \times 3 = 54$.

Quiz for Chapter 4

1. Is the relation {(7, 4) (8, 2) (9, 5) (3, 2)} a function?

   (A) No, because the 2 is repeated.
   (B) Yes, because the 2 is repeated
   (C) No, because each input has exactly one output.
   (D) Yes, because each input has exactly one output.

2. Is the following input-output table a function?

Input	Output
−1	0
0	1
1	3
5	3
1	2

   (A) Yes, the input-output table is a function because there are exactly five input values and five output values.
   (B) No, the input-output table is not a function because the input value of 1 has two distinct output values.
   (C) Yes, because no pair of input-output values contains identical numbers.
   (D) No, because the number zero occurs as both an input and output value.

3. Which of the following has a domain equal to {7, 8, 9, 10} and a range of {2, 3, 4, 5}?

   (A) {(7, 5) (8, 4) (9, 3) (10, 2)}
   (B) {(7, 2), (9, 3), (10, 4), (5, 8}
   (C) {(7, 4), (10, 5), (3, 9), (2, 8)}
   (D) {(7, 3), (4, 8), (2, 10), (5, 9)}

4. Which test can be used to determine if a graph represents a function?

   (A) Vertical line test
   (B) Horizontal line test
   (C) Diagonal line test
   (D) Any straight line test

5. Which of the following graphs does *not* express a function?

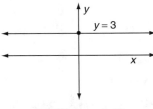

(A) Figure 4.10

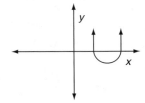

(B) Figure 4.11

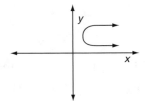

(C) Figure 4.12

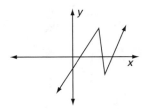

(D) Figure 4.13

6. Christina thinks she can run faster if she sleeps more hours the previous night. She created an input-output table with hours slept as the input and her times in the 100-meter dash as her output.

Input: Hours slept	Output: Time in seconds
6	14.2
6.6	14.1
7	13.8
8.5	11.8
8	12.3

What conclusion can she make regarding the data?

(A)   There is no relationship between the number of hours of sleep and her running time.

(B)   As the number of hours of sleep increases, her running time decreases.

(C)   As the number of hours of sleep increases, her running time increases.

(D)   For every additional hour of sleep, her running time changes by 0.4 seconds.

7.   A manufacturer of duct tape compiles data regarding the strength of its tape. The company carries tape with different widths: 1 inch, 2 inches, 2.5 inches and 4 inches. An experiment was designed to see how much pressure each of the different tapes could withstand before breaking. The pressure was measured in pounds per square inch.

Input: Tape width in inches	Output: Pressure in pounds per square inch
1	2.5
2	5
2.5	6.25
4	10

The company is interested in manufacturing a tape that is 5 inches wide. Based on this table, how much pressure (pounds per square inch = psi) could it withstand ?

(A)   8 psi

(B)   10 psi

(C)   11.5 psi

(D)   12.5 psi

Answer Key

1. (D) The relation is a function because each input has exactly one output. A relation can still be a function if different inputs have the same output.

2. (B) The input-output table does not represent a function because the input, 1, has two outputs, 3 and 2.

3. (A) For answer choice (A), the domain is the set of all inputs, which is {7, 8, 9, 10}. The range is the set of all outputs, which is {5, 4, 3, 2}.

4. (A) A vertical line test shows whether or not a graph represents a function. In order for a graph to represent a function, a vertical line can intersect the graph at only one point.

5. (C) Answer choice (C) is not a function because it does not pass the vertical line test. Vertical lines intersect the graph at more than one point.

6. (B) Christina can conclude that the more hours she sleeps the previous night, the shorter, or *faster*, her running time will be.

7. (D) When you multiply each input by 2.5, you get the output shown in the table. For example, $1 \times 2.5 = 2.5$ and $2.5 \times 2.5 = 6.25$. Therefore, a tape that is 5 inches wide should be able to withstand 12.5 psi : $5 \times 2.5 = 12.5$.

Chapter 5: Algebra—Part 3

Welcome to Chapter Five. In this chapter, we will discuss linear functions. Your goals for this chapter are to:

a) Learn how to distinguish a linear function from other functions.

b) Learn how to express a linear function in an input-output table

c) Learn how to express a linear function graphically

d) Use linear functions to model situations from daily life.

Linear Functions

A linear function, when graphed, lies along a straight line. We have already seen examples of linear functions in Chapter 4. Let's revisit Example 6 from that chapter the question relating holes and shoelaces. First, let's look at the input-output table again, this time with a few added values:

Input	Output
Holes	Length
2	6
3	9
4	12
6	18
8	24

Let's graph the inputs and outputs, letting inputs be the x-coordinate and outputs the y-coordinate.

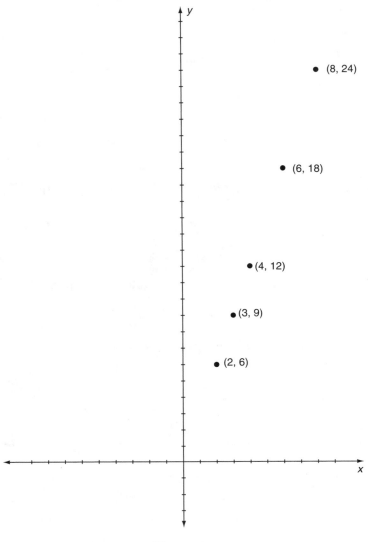

Figure 5.1

Since each *y*-coordinate is triple the *x*-coordinate, we express the function as $y = 3x$.

We can now predict any length of shoelace, given the number of holes.

Try it!

Example 1:

The Shocking Shoe Company needs to buy shoelaces for its next shipment of running shoes. The company purchased 15 pairs of shoes, each with 8 holes. Use the linear function $y = 3x$, for which *y* represents the number of inches and *x* represents the number of holes, to determine how many inches of shoelace are required.

Solution:

Determine the number of holes in 15 pairs of shoes:

$15 \times 2 \times 8 = 240$

Multiply 240 by 3 to find the total length of shoelace needed for this order:

$240 \times 3 = 720$

The Shocking Shoe Company will need 720 inches of shoelace for its next order.

Let's take another look at Figure 5.1, this time focusing on the first three ordered pairs:

$(2, 6), (3, 9), (4, 12)$.

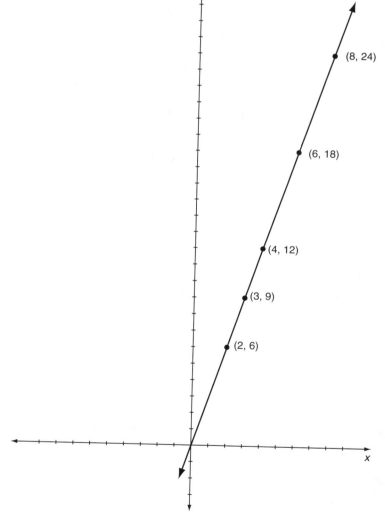

Figure 5.2

As the value of x moves one unit to the right, y increases three units vertically. The value of 3 in the equation $y = 3x$ is called the slope of the graph of the equation, where slope is defined as $\dfrac{rise}{run}$.

Example 2:

What is the slope of the following graph?

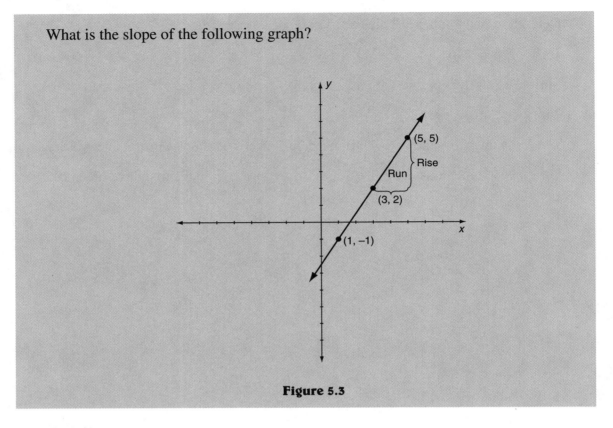

Figure 5.3

Solution:

The slope of the graph is $\dfrac{3}{2}$. Look at the points (3, 2) and (5, 5). As the x coordinate increases 2 units, from 3 to 5, the y coordinate increases 3 units, from 2 to 5. In terms of the slope, the graph rises 3 units and runs 2 units. Thus the slope is $\dfrac{3}{2}$.

If you know the value of the slope, you can graph linear functions without resorting to an input-output table.

Example 3:

Starting from the point (0, −3), graph the next two integer points (as x increases) using a slope of $\dfrac{2}{3}$.

Solution:

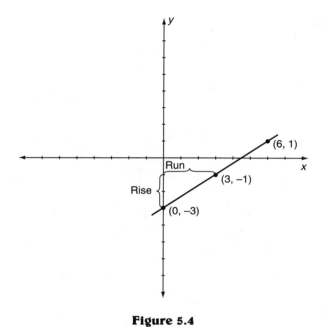

Figure 5.4

By rising 2 and running 3, we arrive at the point (3, −1). By repeating the process, we arrive at (6, 1).

Linear Functions as Models

Linear functions can be used to model situations in our lives.

Try it!

Example 4:

Sean makes extra money by word processing pages for a fee. He charges $3.00 per page. The following table represents his earnings this week.

x Pages	Y $ earned
4	12
2	6
3	9

The following is a graph of the function.

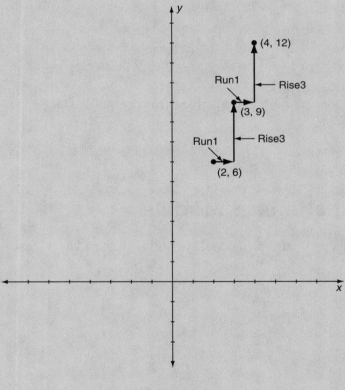

Figure 5.5

a)   Do the table and the graph express a linear function?

b)   What is the slope of the function?

c)   What equation models this function?

d)   How much would Sean earn if he word processed a document that was 17 pages long?

Solution:

a) Both the table and the graph express a linear function. The points lie along a straight line.

b) Look at the points (2, 6) and (3, 9). As the x-value increases by 1 unit (run), the y-value increases by 3 units. (rise) Therefore, $\dfrac{rise}{run}$ is $\dfrac{3}{1}$ or 3. The slope of this linear function is 3.

c) The equation of this linear function is $y = 3x$

d) Substituting 17 for the x-value in the equation $y = 3x$, we calculate that Sean would earn $51 for the job.

Non-linear Functions

As stated earlier, linear functions, when graphed, form a straight line. But not all functions are linear. Let's look at two examples.

Try it!

Example 5:

Create an input-output table for the equation $y = x^2$. Then graph the equation.

Solution:

x	y
-2	4
-1	1
0	0
1	1
2	4

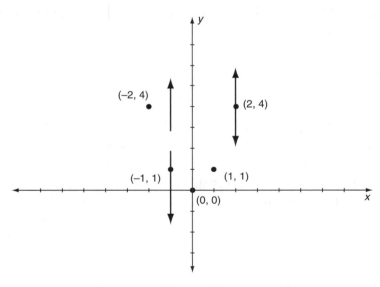

Figure 5.6

Each input (x) has exactly one output (y). Further, a few vertical lines would confirm the graph is a function. However, it does not form a straight line and is therefore a non-linear function.

Example 6:

Graph the points $(-1, 5)$ $(3, -2)$ $(5, 0)$ $(0, -3)$. Is the graph a function? If yes, is it a linear function?

Solution:

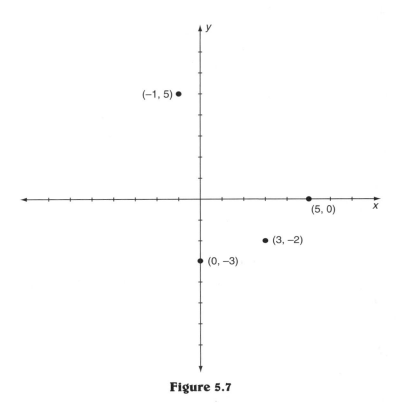

Figure 5.7

Each input has exactly one output, so the graph expresses a function. Since the points do not lie along a straight line, the graph does not express a linear function.

Quiz for Chapter 5

1. Which of the following groups of points represents a linear function?

   (A) $(-2, 4)$, $(0, 0)$, $(3, -6)$, and $(1, -2)$
   (B) $(-2, 4)$, $(1, 8)$, $(4, 12)$, $(7, 14)$
   (C) $(-2, 4)$, $(2, 7)$, $(6, 12)$, $(-6, 1)$
   (D) $(-2, 4)$, $(3, -1)$, $(-7, -11)$, $(9, -6)$

2. What would be the equation for the function shown in question 1?

   (A) $y = 2x$
   (B) $y = -2x$
   (C) $y = \dfrac{1}{2}x$

   (D) $y = -\dfrac{1}{2}x$

3. A Junior Theater group realized that the linear function for its ticket sales was $y = 5.5x$, where x represents the ticket sales and y represents the money earned from the ticket sales. If the theater group sold 420 tickets for the opening night of a play, how much money would the troupe earn?

   (A) $1,424
   (B) $1,928
   (C) $2,310
   (D) $2,500

4. If the same Junior Theater group found it had earned $3,971 on the second evening of its performance, how many tickets were sold?

   (A) 722
   (B) 604
   (C) 588
   (D) 421

5. What is the slope for the linear expression expressed in the following input (x)-output (y) table?

x	y
0	0
1	-2
2	-4
3	-6

(A) $\dfrac{1}{2}$

(B) 2

(C) -2

(D) $-\dfrac{1}{2}$

6. Slope can be defined by what expression?

(A) $\dfrac{rise}{run}$

(B) $\dfrac{run}{rise}$

(C) run $\times$ rise

(D) run $+$ rise

7. Starting from the point $(1, -1)$, use the slope $\dfrac{4}{3}$ to determine the next two points (as x increases) on the graph.

(A) $(4, -5), (7, -9)$

(B) $(5, -4), (9, -7)$

(C) $(5, 2), (9, 5)$

(D) $(4, 3), (7, 7)$

Answer Key

1. (A) The graph expresses a linear function because all of the points lie along a straight line.

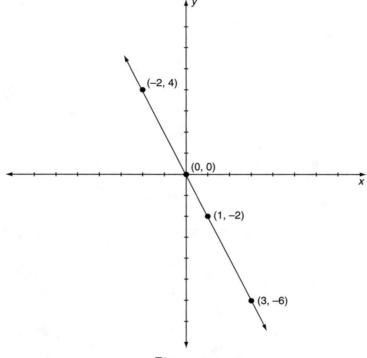

Figure 5.8

2. (B) Each input, when multiplied by -2, gives the output in the table. Therefore, the equation for the linear function is $y = -2x$.

3. (C) Substitute 420 for x, the number of tickets sold. Then $y = (5.5)(420) = \$2310$.

4. (A) Substitute 3971 for y, the earnings from tickets sold, to find x, the number of tickets sold. Then $3971 = 5.5x$. Thus, $x = \dfrac{3971}{5.5} = 722$.

5. (C) Each input is multiplied by -2 to get the output. Therefore, the slope of the linear function is -2.

6. (A) $\dfrac{rise}{run}$ expresses the slope of a linear function.

7.   (D)

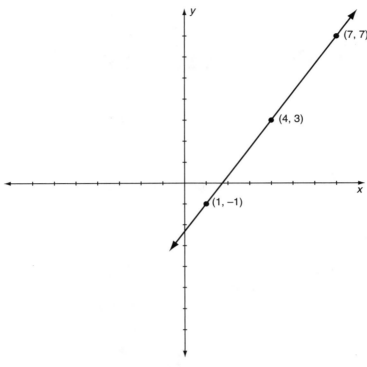

Figure 5.9

Chapter 6: Algebra—Part 4

Welcome to Chapter 6. Your goals for this chapter are to:

(1) Learn about the slope-intercept form of a line.

(2) Learn about the standard form of a line.

(3) Graph pairs of linear equations.

(4) Graph pairs of linear inequalities.

Slope-Intercept Form of a Line

In Chapter 5, we graphed linear functions using the slope and a point on the line.

In this section, we will learn about a special point called the y-intercept.

Any non-vertical line on a graph can be expressed by using the slope-intercept form of a line. The general equation of a line in slope-intercept form is $y = mx + b$,

where m is the slope and b is the y-intercept. In the graph below, the y-intercept is 5 and the slope is $\frac{-2}{5}$.

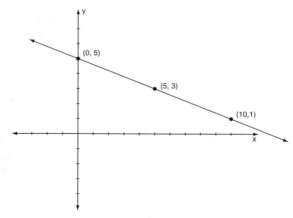

Figure 6.1

The equation of this line, in slope-intercept form, is $y = \dfrac{-2}{5}x + 5$.

Try it!

Example 1:

In the equation $y = 4x - 7$, what are the slope and the y-intercept?

Solution:

By substitution, $m = 4$ and $b = -7$. Thus, The slope is 4 and the y-intercept is –7.

Example 2:

Graph the line $y = 4x - 7$.

Solution:

From the y-intercept of $(0, -7)$, go up 4 units and move 1 unit to the right. Note that the slope, 4, should be thought of as $\dfrac{4}{1}$ when graphing.

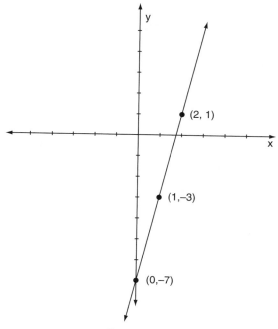

Figure 6.2

Applications of the Formula $y = mx + b$

The slope intercept form of a line has real applications.

Try it!

Example 3:

The fee schedule for a taxicab company is as follows:

$3 boarding fee and $2 per mile traveled

(a) What linear equation models the fee schedule?

(b) What is the cost of a 5-mile trip?

(c) Graph the linear function that models the company's fee schedule.

Solutions:

(a) Let x = miles traveled and y = cost. At the start of the trip, y = 3 and x = 0. The boarding fee, $3, is charged for all trips. Therefore, the y-intercept for the graph is 3. Since $2 is charged per mile traveled, the slope is 2. The equation that models the company's fee schedule is $y = 2x + 3$.

(b) To find the cost of a 5-mile trip, replace x with 5 in the equation.

$y = (2)(5) + 3 = 13$. Thus, the cost of a 5 mile trip is $13.

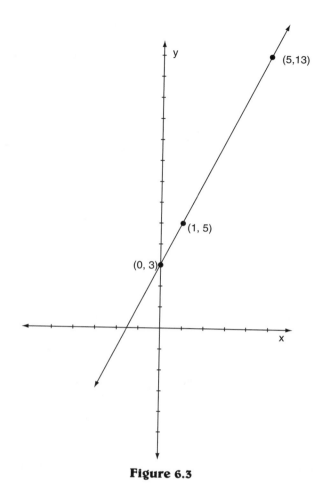

Figure 6.3

The model cannot represent negative values for *x*, because miles traveled must be positive.

The Standard Form of a Line

Linear functions can be expressed either in the slope-intercept form $y = mx + b$ or in the standard form $ax + by = c$ (where *a*, *b*, and *c* must be integers). Many texts prefer that *a* represent a positive integer. Note that even though both forms use the letter *b*, the meaning of *b* is different. In the form $y = mx + b$, *b* represents the y-intercept. In the form $ax + by = c$, *b* simply represents the coefficient of *y*.

Try it!

Example 4:

What is the standard form of the equation $4x - \dfrac{3}{2}y = -7$?

Solution:

Multiply the equation by 2 to eliminate the fraction.

$2\left(4x - \dfrac{3}{2}y = -7\right)$. Then $8x - 3y = -14$.

Example 6:

What is the standard form of the equation $y = \dfrac{-2}{3}x + 8$?

Solution:

Multiply the equation by 3. So $(3)(y) = (3)\left(\dfrac{-2}{3}x + 8\right)$, which becomes $3y = -2x + 24$.
Finally, add $2x$ to each side to get $2x + 3y = 24$.

Example 7:

What is the slope-intercept form of the equation $7x - 5y = 13$?

Solution:

Isolate $-5y$ by subtracting $7x$ from both sides of the equation. Then $-5y = -7x + 13$.
Now divide both sides by -5 to get the final answer of $y = \dfrac{7}{5}x - \dfrac{13}{5}$.

Graphing Lines in Standard Form

Lines in standard form can be easily graphed by using x- and y-intercepts. You can find the x-intercept by replacing y with 0. You can find the y-intercept by replacing x with 0.

Try it!

Example 8:

Graph the equation $4x - 3y = 12$ by using x- and y-intercepts.

Solution:

Find the x-intercept by replacing y with 0. Then $4x - (3)(0) = 12$. So $4x = 12$. Dividing both sides by 4 yields $x = 3$. Thus, the coordinates for the x-intercept are $(3,0)$.

Find the y-intercept by replacing x with 0. Then $(4)(0) - 3y = 12$. So $-3y = 12$. Dividing both sides by -3 yields $y = -4$. Thus, the coordinates for the y-intercept are $(0,-4)$.

Now, graph the x- and y-intercepts by connecting the points $(3, 0)$ and $(0, -4)$ with a line.

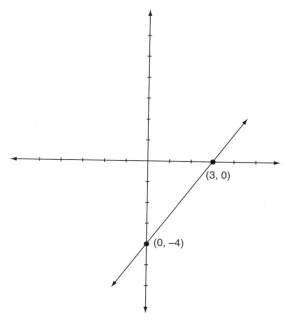

Figure 6.4

Linear Inequalities

In Chapter 3, we learned how to graph the inequality $x \geq -7$ on a number line. We can now use our knowledge of linear functions to graph a linear inequality.

Try it!

Example 9:

Graph $y \leq \dfrac{-2}{3}x + 5$ on a coordinate plane.

Solution:

First, graph $y = \dfrac{-2}{3}x + 5$. Start with the y-intercept $(0, 5)$. Since the slope is $-\dfrac{2}{3}$, we can locate the next point by counting 2 units down (for the new y value) and 3 units to the right (for the new x value). Thus, this next point is $(3, 3)$. Repeating this procedure, we arrive at the point $(6,1)$.

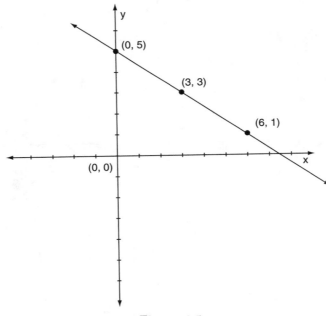

Figure 6.5

Since the graph represents a linear inequality, there will be several values that satisfy the expression. The easiest way to identify those values is to substitute the coordinates of a point on the graph to see if it satisfies the inequality. It is often useful to use the point $(0, 0)$ to minimize calculations. Then $0 \leq \dfrac{-2}{3}(0) + 5$, which leads to $0 \leq 5$.

Since the statement $0 \leq 5$ is a true statement, all values on the (0, 0) side of the line are part of the solution set.

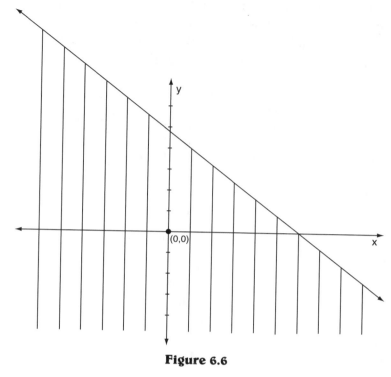

Figure 6.6

Had the line actually passed through (0, 0), we could have used any other point on the graph to test for solutions.

Observe that the line in the previous inequality appeared as any of the lines we have graphed so far. However, sometimes we need to use a dashed (or dotted) line to graph an inequality. A dashed line shows that the line is not part of the graph.

Try it!

Example 10:

Graph the inequality $y > 4x - 2$.

Solution:

We are given that y is greater than $4x - 2$, but it is not greater than or equal to $4x - 2$. Because the line is not included in the inequality, we graph it as dotted or dashed. In some ways, this reminds us of graphing an inequality such as $x > -2$. Since x is greater than -2 but not greater than or equal to -2, we graph the inequality with an open circle at -2. Similarly, in a linear inequality, we show that the line is not part of the graph by using a dashed or dotted line. Graph the line by starting at $(0, -2)$ and use the slope of 4, to find other points on the graph. Similar to Example 9, we find the next point by adding four units to the y value and one unit to the x value of $(0, -2)$. Thus, the next point becomes $(1, 2)$. Repeating this procedure, our subsequent point becomes $(2, 6)$. Now draw a dashed line.

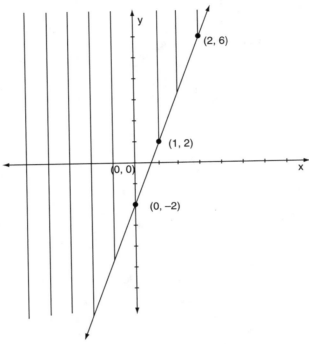

Figure 6.7

Test the inequality by substituting $(0, 0)$ for x and y. So $0 > (4)(0) -2$, which leads to $0 > -2$, which is true. Substituting $(0, 0)$ provided a true statement in the inequality, so shade the side of the graph that contains the origin.

Applications of Linear Inequalities

Linear inequalities can be used to model real-life situations.

Try it!

Example 11:

A local gym club charges \$30 for a membership fee and then \$20 per month for dues. A competing club boasts of beating those prices. Graph an inequality that represents the price structure for the competitor.

Solution:

The first gym charges $30 for membership and $20 per month for dues, so the linear function for the cost of its service is $y = 20x + 30$. Since the competing club offers a better price, their linear inequality is $y < 20x + 30$. The graph of the inequality is shown below.

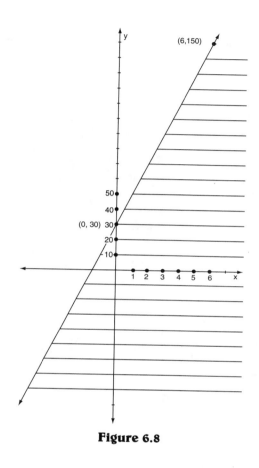

Figure 6.8

If a customer wanted to join the second club for 6 months, we can see from the graph that he/she would pay less than $150. By substitution, we get $y < 20(6) + 30$, which simplifies to $y < 150$.

Systems of Equations

A system of equations is a set of two or more equations with two or more common variables. Systems of equations can be solved algebraically or by graphing.

When solving a system of equations algebraically, we can use the substitution method or the linear combination method (also known as the addition-subtraction method or the elimination method).

Try it!

Example 12:

Solve the system of equations by using the substitution method.

$y = x + 1$

$2x + y = 4$

Solution:

Using the first equation, substitute the expression $x + 1$ for y into the second equation. Then $2x + (x+1) = 4$, which becomes $3x + 1 = 4$. Subtracting 1 from each side, we get $3x = 3$. Divide both sides by 3 to get $x = 1$.

Now solve for y by inputting 1 for x into either equation. Using the first equation, we get $y = 1 + 1 = 2$. (It doesn't get any easier!!) Thus, the solution set for the system of equations is $(1, 2)$.

We can solve the same system of equations by using the linear combination method. First, we must make sure that both equations are in standard form ($ax + by = c$). For the first equation, $y = x + 1$, subtract x from each side to get $y - x = 1$. Then rearrange the terms so that the equation reads as $-x + y = 1$. Now multiply by -1 so that this equation becomes $x - y = -1$.

The second equation, $2x + y = 4$, is already in standard form. Now, place both equations, one over the other, and add vertically.

$x - y + -1$
$+$
$2x + y = 4$
_____
$3x = 3$
$\dfrac{3x}{3} = \dfrac{3}{3}$
$x = 1$

Substitute 1 for x into either original equation. If we use the second equation, $(2)(1) + y = 4$. Then $2 + y = 4$. Subtracting 2 from each side yields the answer $y = 2$. We arrive at the same solution as we did by using the substitution method, namely $(1, 2)$.

Graphing Systems of Equations

A system of equations can also be solved by using graphing. Let's graph both equations from Example 12, leaving each in its original form.

Graph $y = x + 1$ by starting at $(0, 1)$ and counting one to the right and one unit up. So, we arrive at the point $(1, 2)$. Draw the line that connects $(0, 1)$ and $(1, 2)$.

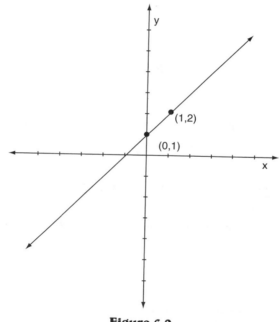

Figure 6.9

Next graph the equation $2x + y = 4$ by using x-and y-intercepts.

To find the x-intercept, let $y = 0$. Then $2x + 0 = 4$. Rewrite as $2x = 4$ and divide both sides by 2. Thus $x = 2$. This yields the point $(2, 0)$.

To find the y-intercept, let $x = 0$. Then $(2)(0) + y = 4$, which means that $y = 4$. This yields the point $(0, 4)$.

Draw the line that connects $(2, 0)$ and $(0, 4)$ on the same graph as $y = x + 1$.

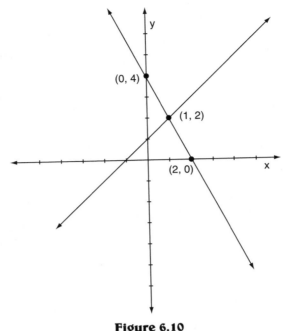

Figure 6.10

Notice that the lines intersect at (1, 2), in agreement with the answers found by using the substitution and linear combination methods.

Graphing Systems of Inequalities

Using our knowledge of systems of equations, we can now graph systems of inequalities.

Try it!

Example 13:

Graph the system of inequalities. $y < 2x - 1$

$y \geq -3x + 5$

Solution:

Using the techniques we learned earlier in the chapter, graph the first inequality, $y < 2x - 1$. Remember, the line in this inequality will be dashed. Test the inequality by substituting (0, 0) for x and y. Then $0 < (2)(0) - 1$, which leads to $0 < -1$. The statement $0 < -1$ is false, so shade the other side of the dashed line. (Some students find it easier to simply input arrows, in order to remind them where to shade at the end of the problem.)

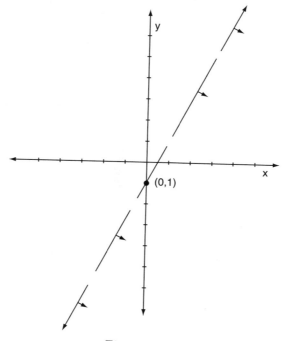

Figure 6.11

Next, graph the inequality $y \geq -3x + 5$. This time the line will be solid because the inequality includes the equals sign.

Test this inequality by substituting (0, 0) for x and y. Then $0 \geq (-3)(0) + 5$, which becomes $0 \geq 5$.

The statement, $0 \geq 5$, is false so shade the other side of this line.

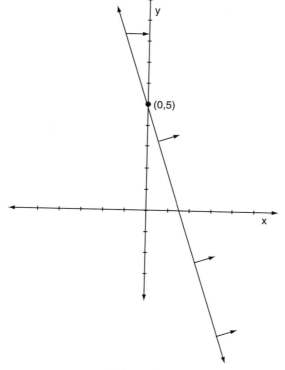

Figure 6.12

The solution to this set of inequalities is the common area shaded in the final graph for both inequalities, as shown below.

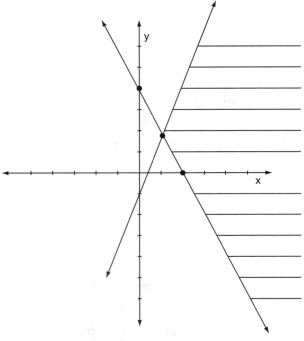

Figure 6.13

Quiz for Chapter 6

1. An equation of a line in the form of $y = mx + b$ is called _____ form.

    (A) standard
    (B) slope-intercept
    (C) inequality
    (D) system

2. Which of the following describes correctly both the slope and y-intercept of the graph of the equation $y = -4x + 3$?

    (A) Slope is 1, y-intercept is 1
    (B) Slope is 3, y-intercept is -4
    (C) Slope is -4, y-intercept is 3
    (D) Slope is -3, no y-intercept

3. A car rental agency charges a rental fee of $30 plus $0.60 per mile traveled. What linear function, in slope-intercept form, models this fee schedule?

    (A) $y = 60x + 30$
    (B) $y = 60x - 30$
    (C) $y = -0.60x + 30$
    (D) $y = 0.60x + 30$

4. Using the fee schedule for the car rental agency in question #3, what would be the cost of driving 80 miles?

    (A) $30
    (B) $48
    (C) $78
    (D) $80

5. Look at the following graph of line l.

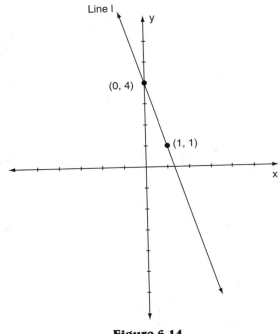

Figure 6.14

What is the equation of this line?

(A) $y = 3x + 4$
(B) $y = -3x + 4$
(C) $y = 3x - 4$
(D) $y = -3x - 4$

6. The equation $4x - 2y = 8$ is a line in what form?

(A) Standard form
(B) Slope-intercept form
(C) Standard-intercept form
(D) Inequality form

7.   Look at the following graph of line *m*.

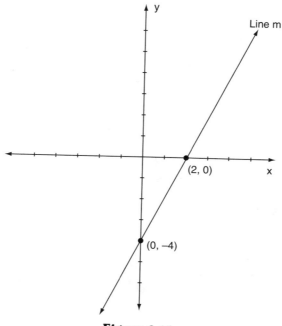

Figure 6.15

Which of the following is completely correct?

(A)   The slope is 2 and the *x*-intercept is −4.
(B)   The slope is −2 and the *x*-intercept is −4.
(C)   The slope is 2 and the *x*-intercept is 2.
(D)   The slope is −2 and the *x*-intercept is 2.

8.   Which of the following is the standard form of the equation $y = \dfrac{-3x}{2} + 2$?

(A) $2x + 3y = 4$
(B) $4x - 3y = 6$
(C) $3x + 2y = 4$
(D) $x + y = 6$

9. Look at the following graph.

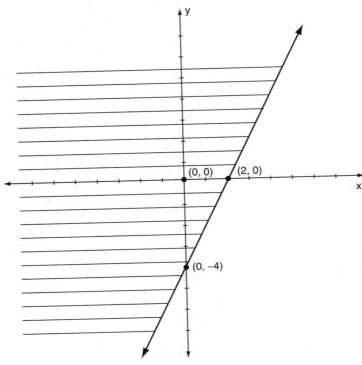

Figure 6.16

Which inequality is shown by the shaded region?

(A) $y \leq 2x - 4$
(B) $y < 2x - 4$
(C) $y \geq 2x - 4$
(D) $y > 2x - 4$

10. What is the solution of the following system of equations?

$2x - y = 8$
$x + y = 1$

(A) $(3, -2)$
(B) $(2, -3)$
(C) $(-3, 2)$
(D) $(-2, -3)$

11. Consider the following system of equations.

$3x + y = 8$
$y = 2x - 2$

The substitution method will be used to first find the value of x. Which of the following describes a correct procedure?

(A)   Determine that $y = 3x + 8$ from the first equation and substitute this expression for y in the second equation.

(B)   Determine that $y = -3x + 8$ from the first equation and substitute this expression for y in the second equation.

(C)   Rewrite both equations in slope-intercept form.

(D)   Rewrite both equations in standard form.

12. Look at the following graph.

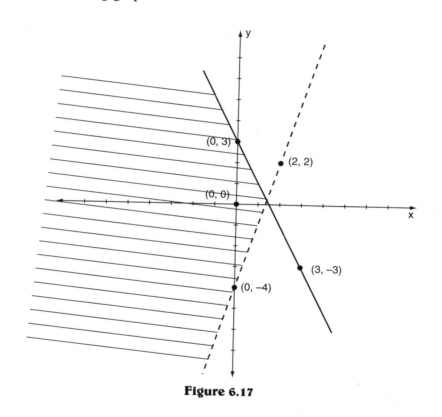

Figure 6.17

Which system of inequalities is illustrated?

(A) $y < 3x - 4$

   $y \geq -2x + 3$

(B) $y \geq 3x - 4$

   $y < -2x + 3$

(C) $y \leq 3x - 4$

   $y < -2x + 3$

(D) $y > 3x - 4$

   $y \leq -2x + 3$

Answer Key

1. (B) The equation $y = mx + b$ represents the slope-intercept form of a line.

2. (C) In the form $y = mx + b$, m represents the slope and b represents the y-intercept.

3. (D) The fixed cost, \$30, is the y-intercept. Since the final cost varies with the number of miles traveled, 0.60 represents the slope.

4. (C) Using the formula $y = 0.60x + 30$, substitute 80 for x. Then $y = (0.60)(80) + 30 = 48 + 30 = 78$.

5. (B) In moving from (0, 4) to (1,1), the change in y is –3 units and the change in x is 1 unit. So the slope equals $\frac{-3}{1} = -3$. The point (0, 4) reveals that the y-intercept is 4. Thus, the equation is $y = -3x + 4$.

6. (A) Standard form is the form of $ax + by = c$, where a, b, and c are integers. For the equation $4x - 2y = 8$, $a = 4$, $b = -2$, and $c = 8$.

7. (C) In moving from (0, –4) to (2, 0), the change in y is 4 units and the change in x is 2 units. So the slope equals $\frac{4}{2} = 2$. The point (2, 0) reveals that the x-intercept is 2.

8. (C) Multiply both sides of the equation by 2. Then $2y = (2)\left(\frac{-3x}{2}\right) + (2)(2)$, which becomes $2y = -3x + 4$. Finally, add $3x$ to each side to get $3x + 2y = 4$.

9. (D) We graph the line $y = 2x - 4$ by using the points (0, 4) and (2, 0). Since the shaded area lies above the dashed line, the answer is $y > 2x - 4$.

10. (A) Add the two equations.

$$\begin{array}{r} 2x - y = 8 \\ + \quad x + y = 1 \\ \hline 3x = 9 \end{array}$$

$$\frac{3x}{3} = \frac{9}{3}, \quad \text{so } x = 3.$$

Substitute $x = 3$ into the second equation to find y.

$3 + y = 1$, so $y = -2$.

11. (B) Solve the first equation for y to get $y = -3x + 8$. This expression will then be used for y in the second equation.

The second equation will actually read as $-3x + 8 = 2x - 2$. The solution is $(2, 2)$.

12. (D) For the dashed line, the slope is 3 and the y-intercept is -4. So, its equation is $y = 3x - 4$. For the solid line, the slope is -2 and the y-intercept is 3. So, its equation is $y = -2x + 3$. The shaded area lies above the dashed line, but below the solid line.

Thus $\begin{array}{l} y > 3x - 4 \\ y \leq -2x + 3 \end{array}$ represents the correct system of inequalities.

Chapter 7: Data Analysis and Probability

Welcome to Chapter Seven. Your goals for this chapter are to:

a)   Learn about sets and Venn diagrams.

b)   Learn about the Counting Principle

c)   Learn about Probability.

d)   Determine a line of best fit.

Sets and Venn Diagrams

Consider all the whole number factors of 24 and 15.

24: 1, 2, 3, 4, 6, 8, 12, 24

15: 1, 3, 5, 15

If you were asked to find the common whole number factors of both 24 and 15, you could express the information using a Venn diagram. A Venn diagram helps us picture numerical relationships. In the diagram below:

A shows the whole number factors of 24.

B shows the whole number factors of 15.

W represents the group of all whole numbers.

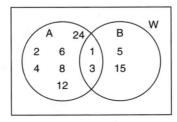

Figure 7.1

W is also called the universal set.

This Venn diagram has helped us see that the common factors of 15 and 24 are 1 and 3.

The same problem can be explained by using sets. A set is simply any collection of objects. For example, the fruits sold at a fruit stand are a set. We could write this set as: $A = \{$apple, orange, banana, pineapple, mango$\}$.

This would be interpreted to mean that A is the set of fruits sold at a particular fruit stand.

Each fruit is called an element of the set. To demonstrate that an apple, for example, is an element of A, we would write "apple $\in A$", which would be read as "apple is an element of A."

Try it!

Example 1:

Use set notation to show the set of all whole number factors of 24. Create another set for all the whole number factors of 15.

Solution:

$A = \{1, 2, 3, 4, 6, 8, 12, 24\}$ represents the set of all whole number factors of 24.

$B = \{1, 3, 5, 15\}$ represents the set of all whole number factors of 15.

Subsets

A **subset** of a given set is a set that contains some of the elements of the given set. For example $\{2, 3, 4\}$ is a subset of $\{1, 2, 3, 4, 6, 8, 12, 24\}$. Likewise, $\{15\}$ is a subset of $\{1, 3, 5, 15\}$. We use the symbol $\subseteq$ to denote a subset. In Example 2, we will learn about two special subsets.

Try it!

Example 2:

List all of the subsets of $A = \{1, 2, 3\}$.

Solution:

First, list all the subsets with one element. They are $\{1\}$, $\{2\}$, and $\{3\}$.

Second, list all the subsets with two elements. They are {1, 2}, {2, 3}, and {1, 3}. Note, for example, that {2, 1} is equivalent to {1, 2}. For this reason, {2, 1} is not listed. We are looking for all *different* subsets.

Third, we list {1, 2, 3}, since every set is a subset of itself.

There is one last set called the empty set or the null set. A set with no elements is also a subset of every set. It is written as { } or $\varnothing$, but *not* {$\varnothing$}.

Counting the null set and the set A itself, $A = \{1, 2, 3\}$ has eight subsets.

Complements

The set of elements that are not in a set, but are part of the universal set, is called the **complement**. For any set A, the complement would be noted as $\overline{A}$.

Try it!

Example 3:

What is $\overline{A}$ if $A = \{2, 4, 6, 8)$ and U, the universal set, is $\{1, 2, 3, 4, 5, 6, 7, 8\}$?

Solution:

The elements in U that are not in A are 1, 3, 5 and 7. Therefore $\overline{A} = \{1, 3, 5, 7\}$.

Intersection and Union of Sets

In the beginning of this chapter, we showed that the common factors of 15 and 24 are 1 and 3. The subset, {1, 3}, is also the intersection of the two sets. We use the symbol $\cap$ to denote the intersection of two sets.

Try it!

Example 4:

Using set notation, show that the intersection of the set of whole number factors of 24 and 15 is {1, 3}

Solution:

$\{1, 2, 3, 4, 6, 8, 12, 24\} \cap \{1, 3, 5, 15\} = \{1, 3\}$.

A **union of sets**, written as A $\cup$ B, is the set of elements that are in either A or B or both.

Example 5:

Given $S = \{2, 3, 4, 5, 6\}$ and $T = \{3, 4, 5, 6, 7\}$, describe $S \cup T$.

Solution:

$S \cup T = \{2, 3, 4, 5, 6, 7\}$.

The Counting Principle

If a school lunch menu offers two different types of sandwiches, three different side orders and two different beverages, how many different combinations of sandwich, side order and beverage are available? One way we can express the answer is by using a tree diagram.

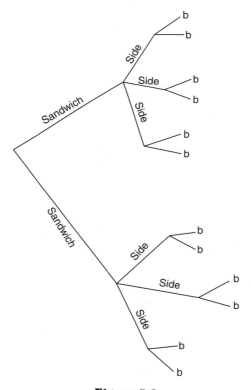

Figure 7.2

The **tree diagram** is a useful visual aid to demonstrate that there are a total of 12 different combinations of sandwich, side order, and beverage. However, we ought not draw a tree diagram for every problem of this sort. An easier to way to calculate the number of combinations of lunch options is to multiply all the possibilities.

2	×	**3**	×	**2**	=	**12**
sandwiches	×	side orders	×	drinks	=	combinations of lunches

What we have just illustrated above is **the multiplication principle of counting**. Let's look at another example of the counting principle.

Try it!

Example 6:

The third-grade teacher, Mrs. Melon, wants to bring an assortment of fruits to the class. In one box, she has bananas, apples and oranges. In a second box, she has mangoes, plums, and peaches. How many different types of fruit does she have?

Solution:

We can portray this problem by using set notation.

$A = \{$bananas, apples, oranges$\}$

$B = \{$mangoes, plums, peaches$\}$

$A \cup B = \{$bananas, apples, oranges, mangoes, plums, peaches$\}$, which means that she has six different types of fruit.

What we have illustrated in this example is called the addition principle of counting.

Note that none of the fruits in one box were found in the other box.

However, sometimes these situations have additional considerations.

Example 7:

Mrs. Melon was in a hurry to get to school, so her daily assortment of fruits got mixed up. In the first box, she had bananas, apples, and oranges.

In the second box, she had mangoes, peaches, apples, and oranges. How many different types of fruit does she have in the two boxes?

Solution:

Notice that Mrs. Melon has duplicated some of the fruits in each box. We can solve this problem by using set notation. Whenever two sets have identical elements, we can find their union (or sum) by using the following formula:

$(A \cup B) = (A) + (B) - (A \cap B).$

Then $\{$banana, apple, orange$\}$ + $\{$mango, peach, apple, orange$\}$ − $\{$apple, orange$\}$ = $\{$banana, apple, orange, mango, peach$\}$. Thus, the answer is five. Note that we subtract $(A \cap B)$ in order to avoid counting a type of fruit twice.

Probability of Independent Events

When you enter a school raffle for a prize, you are experiencing an example of probability. How many tickets did you buy? How many were sold? These questions affect the probability of your winning the raffle. Probability can be expressed by using the formula:

Probability of an event = favorable outcomes ÷ all outcomes

Our discussion in this section will focus on independent events. Two events are independent if the occurrence of one event has no effect on the occurrence of the other event.

Try it!

Example 8:

> Michael purchased a package of baseball trading cards. He knows among the ten cards in the package, there is one Chipper Jones card. What is the probability he will draw the Chipper Jones card on the first draw?

Solution:

The required probability is favorable outcomes ÷ all outcomes $= 1 \div 10 = \dfrac{1}{10}$.

(The answer, $\dfrac{1}{10}$, may also be expressed in decimal form as 0.1.)

We can use our knowledge of probability to calculate the probability of two independent events occurring.

Example 9:

Jamie flips a coin and then rolls a die. What is the probability that the coin will land on heads and the die will land on an even number?

Solution:

Calculate the separate probability of each event occurring and multiply those probabilities.

The probability of a coin landing on heads is $\frac{1}{2}$.

The probability of a die landing on an even number is $\frac{3}{6} = \frac{1}{2}$.

Multiply the probabilities to get $\frac{1}{2} \times \frac{1}{2} = \frac{1}{4}$.

There is a $\frac{1}{4}$ (or 0.25) probability of a coin landing on heads and a die landing on an even number.

Linear Functions and Scatter Plots

In Chapter 5, we learned how to use linear functions to predict future data. For example, our problem relating shoelace length to the number of holes in a shoe helped us determine the length of shoelace needed for a large order of shoes. Sometimes we can determine a useful linear function simply from observing data.

Try it!

Example 10:

A manufacturer of an upscale automobile wants to assess the resale value of its car. The manufacturer conducted a survey of 10 cars and placed the data in an input-output table. The input, x, was the number of miles the car had driven (in thousands) and the output, y, was the resale value of the car (in thousands $). The following chart represents the findings.

Miles Driven (in thousands)	Resale Value (in thousands of dollars)
0	35
5	30
15	22
15	18
25	16
30	10
30	8
40	6
40	5
40	5

The manufacturer wants to know the resale value of a car that has been driven 24,000 miles. How will the manufacturer calculate this amount?

Solution:

Graph the data in the input-output table. This type of graph is called a scatter plot. If the data roughly lie along a straight line, draw the line. This type of line is known as a line of best fit. Not every point will be on this line, but each point should be close to it. Some of these points will lie above this line of best fit, while other points lie below this line.

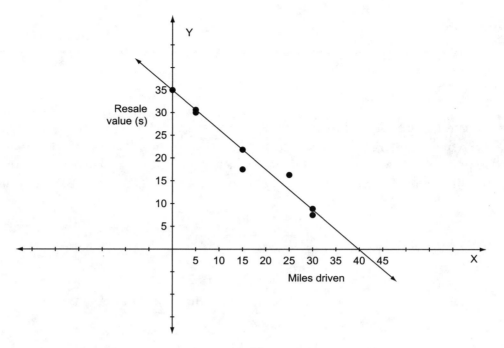

Figure 7.3

We see that the y-intercept of this line is 35. Pick another point on the line that will help you determine the slope. Using the point (30,10), we find $\dfrac{rise}{run}$, from (0, 35) to (30, 10) to be $\dfrac{-25}{30}$ which simplifies to $\dfrac{-5}{6}$. With a slope of $\dfrac{-5}{6}$ and a y-intercept of 35, we find that our linear function for the line of best fit is $y = \dfrac{-5}{6}x + 35$. Calculate the resale value of the car by substituting $x = 24$ (thousand) into the linear function. Then $y = \dfrac{-5}{6}(24) + 35 = 15$.

Thus, the manufacturer estimates the resale value of its car, after having been driven 24,000 miles, is $15,000.

Quiz for Chapter 7

1. In the following Venn diagram, *A* represents the whole number factors of 12 and *B* represents the whole number factors of 21.

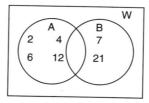

Figure 7.4

Which numbers are missing in the intersection of these two sets?

(A) 12 and 21
(B) 1, 3, 12, and 21
(C) 1 and 3
(D) 12 and 21

2. A member of a set is called a(n) _____.

(A) element
(B) intersection
(C) universe
(D) union

3. Which of the following is a subset of {2, 4, 6, 8, 10, 12, 14}?

(A) {2, 3, 4}
(B) {4, 5, 6}
(C) {2, 5, 14}
(D) {8, 10, 12}

4. What is the proper way to illustrate that set B is a subset of set A?

(A) A ⊆ B
(B) B ⊆ A
(C) A ∅ B
(D) B ∅ A

5. The set D = {a, b, c} has how many subsets?

   (A) 3
   (B) 6
   (C) 8
   (D) 9

6. If A = {2, 3, 4, 6} and the universal set U = {1, 2, 3, 4, 5, 6, 7}, which set describes $\overline{A}$?

   (A) {1, 2, 3, 4, 5, 6, 7}
   (B) {1, 5, 7}
   (C) {2, 4, 6}
   (D) {1, 3, 5, 7}

7. If M = {3, 5, 7, 9} and N = {2, 3, 4, 5}, which set describes $M \cap N$?

   (A) {3, 5}
   (B) {2, 3, 4, 5, 7, 9}
   (C) {2, 3}
   (D) {3}

8. If X = {4, 5, 6, 7, 8} and Y = {6, 7, 8, 9, 10}, which set describes $X \cup Y$?

   (A) {6, 7, 8}
   (B) {4, 5, 6, 7, 8, 9, 10}
   (C) {6, 8}
   (D) {8, 9, 10}

9. Sheila packed four skirts, three blouses and two pairs of shoes for her weekend trip. How many combinations of skirts, blouses, and shoes are possible?

   (A) 9
   (B) 18
   (C) 21
   (D) 24

10. Sammy has four pairs of black socks, three pairs of white socks, and three pairs of brown socks in a drawer. If he reaches into the drawer, what is the probability that he will select a pair of black socks?

(A) $\dfrac{3}{10}$

(B) $\dfrac{2}{5}$

(C) $\dfrac{3}{5}$

(D) $\dfrac{7}{10}$

11. If a coin is flipped three times, what is the probability that it will land on heads all three times?

(A) $\dfrac{1}{8}$

(B) $\dfrac{1}{4}$

(C) $\dfrac{1}{2}$

(D) $\dfrac{3}{4}$

12. A line that lies closest to the points of a scatter plot is called a _____.

(A) curved line
(B) scattered line
(C) line of best fit
(D) line of proportion

13. A coach of a girls' softball team created the following graph. The x value represents the hits each girl had in the game and the y value represents the number of at-bats.

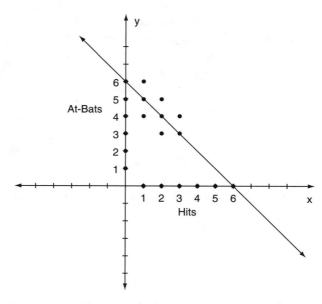

Figure 7.5

What linear function, in slope-intercept form, is the line of best fit?

(A) $y = 2x + 4$
(B) $y = x + 6$
(C) $y = -x + 6$
(D) $y = -x - 6$

14. Given the line of best fit in example 13, if a girl had two hits, how many at-bats would she have had?

(A) 6
(B) 5
(C) 4
(D) 3

Answer Key

1. (C)   The common factors, as expressed through the Venn diagram, are 1 and 3.

2. (A)   By definition, a member of a set is called an element.

3. (D)   The set containing the elements 8, 10, and 12 is also contained in the given set. All the other answer choices have elements that are not contained in the given set.

4. (B)   "B is a subset of A" is written as $B \subseteq A$.

5. (C)   The subsets are: {a}, {b}, {c}, {a, b}, {b, c}, {a, c}, {a, b, c} and $\varnothing$.

Remember that given any set, the set itself and the empty set are subsets.

6. (B) $\overline{A}$, the complement of A, consists of all the elements of the universal set that are not elements of A.

7. (A)   The symbol "$\cap$" means intersection. The intersection of M and N is the set containing the common elements of M and N, which are 3 and 5.

8. (B)   The symbol "$\cup$" means union. $\{X \cup Y\} = \{X\} + \{Y\} - \{X \cap Y\}$. The union of X and Y is the set containing 4, 5, 6, 7, 8, 9, and 10. This set represents all elements that belong to X, to Y, or to both X and Y.

9. (D)   Multiply the number of skirts by the number of blouses by the number of pairs of shoes to find the number of outfit combinations. $4 \times 3 \times 2 = 24$.

10. (B)   To find the probability of selecting a pair of black socks, use the formula: favorable outcomes ÷ total outcomes. Since there are four pairs of black socks, out of a total of ten pairs of socks, the required probability is $\dfrac{4}{10} = \dfrac{2}{5}$.

11. (A)   Each flip of the coin is an independent event because landing on heads on one flip has no effect on future flips. The probability of a coin landing on heads is $\dfrac{1}{2}$. Multiply each probability by the others to find the probability of all three flips landing on heads. Thus, the required probability is $\dfrac{1}{2} \times \dfrac{1}{2} \times \dfrac{1}{2} = \dfrac{1}{8}$.

12. (C)   By definition, a line that lies closest to the points of a scatter plot is called a line of best fit.

13. (C)   When we draw a line of best fit, we can make the following observations: One girl had no hits in her 6 at-bats. The point (0, 6), therefore, is b, the y-intercept. Also, for the points (1, 5) and (2, 4), we notice that $\dfrac{rise}{run} = \dfrac{-1}{1}$. Therefore, the slope of the line, m, is -1.

Thus, the linear function for the scatter plot is $y = -x + 6$.

14. (C)   Substitute 2 for x into the equation $y = -x + 6$. Then $y = -2 + 6 = 4$.

Chapter 8: Mathematical Process Skills

Welcome to Chapter Eight. Your goals for this chapter are to:

(a) Devise and apply alternate strategies for solving problems in mathematics.

(b) Learn about mathematical proofs.

(c) Organize data so valid conclusions can be drawn.

(d) Communicate conclusions based upon data you have organized.

Alternate Strategies for Problem Solving

When you are shopping, it is not always possible to devise equations to assess your costs at the checkout register. Imagine you have purchased two items that, when taxed, cost $9.61 and $4.26. How could you quickly find the sum of your purchases?

Try it!

Example 1:

Karen was at a grocery store and purchased items, including tax, that cost $7.41, $3.81, $5.23, and $6.89. If she has $27 in her purse, how can she quickly conclude whether she can pay for these purchases?

Solution:

Karen can round the cost of her purchases to the nearest dollar to estimate the total cost.

$7.41 rounds to $7.00.

$3.81 rounds to $4.00.

$5.23 rounds to $5.00.

$6.89 rounds to $7.00.

Adding $7, $4, $5, and $7 yields an estimated cost of $23, which is less than the money she has. Thus, she has enough money to pay for her purchases.

Note that Karen could have also used a more conservative approach, rounding each purchase to the next highest dollar.

$7.41 becomes $8.00

$3.81 becomes $4.00

$5.23 becomes $6.00

$6.89 becomes $7.00.

Adding $8, $4, $5, and $7 yields a projected cost of $24, still within Karen's budget. Thus, estimating with different standards is a useful method of solving mathematical problems.

Sometimes, using a guess-and-check strategy is a helpful way to solve problems.

Example 2:

> Milt was given the following math problem in his Algebra class:
> If a person has 10 nickels and dimes worth $.65, how many of each coin does he have?
>
> The teacher suggested a system of equations was necessary and provided the following:
>
> $n + d = 10$
> $.05n + .10d = .65$
>
> Is there another strategy that Milt can use to solve this problem?

Solution:

Given that the number of coins is not large, Milt can use a guess-and-check table to solve this problem. A guess-and-check table uses well-considered guesses to focus on correct solutions. The following is a guess-and-check table that would apply to this problem.

# of nickels	# of dimes	Sum of nickels and dimes	Conclusion
4	6	$4(5) + 6(10) = 80$	Too high
5	5	$5(5) + 5(10) = 75$	Too high
7	3	$7(5) + 3(10) = 65$	Correct Solution

Although the guess-and-check table was a convenient method for solving this question, a good teacher is responsible for teaching students various ways of doing problems. For example, if the quantities were much larger, a system of equations may have been a more efficient way of doing the problem.

Deductive Reasoning

Mathematics is not the only facet of our lives in which logical reasoning is necessary. Have you ever lost your keys and wondered where they were? You could panic and look everywhere, or you could try to remember where you went since the last time you remember having the keys. The problem of the lost keys is an example of deductive reasoning. We draw conclusions of the key's whereabouts by systematically reducing the number of possible locations of the keys.

Inductive Reasoning

Another way to use logic to solve problems is through inductive reasoning. Inductive reasoning is a way of drawing conclusions based on existing information. The conclusion that is reached is called a conjecture.

Try it!

Example 3:

Billy enjoys camping, but he hates the crowds. He has noticed that each weekend, when the predicted temperature exceeded 80°, the campgrounds were overflowing with people. This weekend, the weather is forecast to be 82° to 85° degrees. Should Billy go camping this weekend?

Solution:

Given that Billy hates camping when there are crowds, his conjecture is that he should not go camping this weekend. The temperature, forecast to be over 80°, is sure to bring in a lot of campers.

In Example 3, Billy used inductive reasoning. His knowledge of existing conditions, that high temperatures bring big crowds, helped him make an intelligent decision about his weekend plans.

Proofs

In a court, a lawyer uses existing laws to help prove facts in legal cases. Similarly, in mathematics, we use existing rules and properties to prove conjectures. The format for proving conjectures in mathematics is called a proof. There are several formats for proofs: two-column, paragraph, flow, and indirect. We will limit our discussion to a review of two-column proofs.

Two-Column Proofs

A two-column proof, as its name implies, has two columns. On the left, typically, will be statements that are proposed to prove a conjecture. On the right is a column to provide reasons for the statements made on the left.

The proof begins with certain given conditions. What follows is a conjecture that we need to prove. Let's take a look at an algebraic proof.

Given: $3x + 12 = 24$

Prove: $x = 4$

Statement	Reasons
1. $3x + 12 = 24$	Given
2. $3x = 12$	Subtraction Property
3. $x = 4$	Division Property

There are certain rules that are heeded when creating a two-column proof. Notice, for example, that the first statement is a restatement of the given condition. Further, the reason we provide for the first step is that the condition was given. You might also have noticed that certain steps were omitted. In step #2, we provided "Subtraction Property" as the reason, but did not actually show that the following occurred: $3x + 12 - 12 = 24 - 12$. We simply showed the result, $3x = 12$. Finally, the last statement in a proof is the statement we set out to prove (in this case that $x = 4$). Technically, we could have also shown the third step as $\frac{3x}{3} = \frac{12}{3}$. Then the fourth step would have been $x = 4$.

Organizing Data: The Mean, Median, and Mode

Mean

When you received your last math grade (hopefully an "A"), your teacher compiled your grade by calculating a mean. The **mean** (also known as the average) is the sum of a group of numbers, divided by how many numbers were added.

Try it!

Example 4:

Brianna earned scores of 82, 97, 88, and 73 on her last four tests. What was the mean of these four tests?

Solution:

To find the mean, add up all the scores. We get $82 + 97 + 88 + 73 = 340$. Since Brianna took four tests, divide the sum by 4. Thus, the mean score was $340 \div 4 = 85$.

Median

Another method of interpreting the significance of numbers is to find the median. The **median** of a set of numbers is the one in the middle, when the numbers are arranged in ascending (or descending) order.

Try it!

Example 5:

Brianna's quiz grades were as follows: 88, 66, 91, 43, and 81. What was the median of her quiz grades?

Solution:

To find the median, arrange the numbers from least to greatest. We get 43, 66, 81, 88, and 91.

Since 81 is the quiz grade in the middle, the median for Brianna's quiz grades is 81.

When there are two values in the middle, the median is found by finding the mean of those two numbers.

Example 6:

Brianna earned a 92 on her next quiz. What is the current median of her quiz grades?

Solution:

Return to Example 5 where we arranged the quiz grades in ascending order. Include the new quiz grade, 92, to the list. Now the list reads as 43, 66, 81, 88, 91, and 92.

Notice that there are two quiz scores in the middle, 81 and 88. To find the median of this group, find the mean of 81 and 88. Thus, Brianna's current quiz grade is

$$\frac{81 + 88}{2} = \frac{169}{2} = 84.5.$$

Mode

Identifying the mode of a group of numbers is another useful tool for interpreting data. The **mode** is the number in a list that occurs most frequently. A group of numbers, in which each number occurs only once, has no mode. There may also be more than one mode for a group of numbers.

Try it!

Example 7:

Brianna just took another quiz today and scored 92. What is the mode of her quiz grades?

Solution:

Add 92 to Brianna's list of quiz grades. They now read, in order, as 43, 66, 81, 88, 91, 92, and 92.

Brianna scored 92 on two quizzes. Since 92 is the score that occurs most frequently, the mode of her quiz grades is 92.

A group of numbers can have more than one mode.

Example 8:

Brianna will take her final quiz next week. If her last quiz grade is 91, what will be the mode of all her quiz grades?

Solution:

Add 91 to the list of quiz grades. Her final list of quiz grades, arranged in order, becomes 43, 66, 81, 88, 91, 91, 92, and 92.

Since 91 and 92 both occur two times, there are two modes for this list. The modes are 91 and 92.

We can use our knowledge of the mean, median, and mode to draw conclusions about groups of numbers. We will use a somewhat extreme example to make this point.

Example 9:

The weekly allowance for seven students is as follows:

Student 1: $4, Student 2: $8, Student 3: $10, Student 4: $5, Student 5: $3, Student 6: $100, Student 7: $10

Identify the mean, median, and mode of the weekly allowances. Evaluate the effectiveness of each number in interpreting the list.

Solution:

Calculate the mean, median and mode.

The mean is $\dfrac{4+8+10+5+3+100+10}{7} = \dfrac{140}{7} = \$20$.

To find the median, we rearrange the numbers as 3, 4, 5, 8, 10, 10, and 100.

Thus, the median for this group of numbers is the fourth number, which is $8.

The mode is $10, since it occurs more often than any of the other numbers.

Both the median and the mode are useful indicators of the student's allowances because they appear within the cluster of the numbers. The mean, $20, is larger than all of the values except for $100. It therefore is not a good indicator of the array of the allowances in this group. The value $100 is called an outlier and has exaggerated, or skewed, the data.

Mathematical Applications in Real Life

Mathematics can be used to create well-considered conjectures in everyday life. Advances in health care and technology, just to name two examples, are a direct result of interpreting collections of assembled data. However, we can show how applying mathematics recently helped a middle school get a new swimming pool.

A large parcel of land at a middle school was available for development. The community had discussed projects such as a museum, an additional school lunchroom and a technology center. Two students, however, had different ideas. Gavin and Tanya knew the students wanted a new swimming pool at the school. The land available for development was just the right size for a pool, but the two needed to convince the community that this was the right project. They both reasoned that the students were the ones who used the school, so the students' concerns should be given strong consideration. Gavin and Tanya decided to survey all 800 of the students at the school regarding the use of the land. Gavin would poll those 400 students whose last names began with letters from A to M and Tanya would interview the rest. Together, they devised the following survey:

Please select the land-use project that would most benefit our school:

Technology Center

Additional Lunch Room

Swimming Pool

Museum

Tanya's results were as follows:

Technology Center: 78

Additional Lunch Room: 74

Swimming Pool: 224

Museum: 24

Gavin's results were as follows:

Technology Center: 69

Additional Lunch Room: 77

Swimming Pool: 235

Museum: 19

Although the results were not identical, they were very similar. Gavin and Tanya knew they had clearly discovered, independently, that the students wanted a pool.

The two students were scheduled to speak before the school board the following week, but they did not agree on how to convey their findings. Gavin wanted to compile a simple chart that showed their findings. He felt that if each board member had a copy, the numbers would speak for themselves. Tanya, however, was more ambitious. She suggested the two create a PowerPoint presentation with the following bar chart.

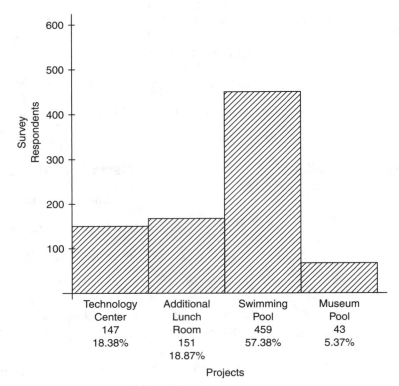

Figure 8.1

After viewing Tanya's bar chart, Gavin agreed it was the most convincing way to influence the board members. The board members agreed, too; the pool was selected to be built the following year.

Tanya and Gavin were successful in convincing the school board to select a pool for the school because they provided an intelligent conjecture, organized the survey results, and used powerful visual tools to make their case.

Quiz for Chapter 8

1. Emily was purchasing three blouses that cost, including tax, $12.49, $13.29 and $19.89. Without using any paper, pencil, or a calculator, how can she quickly assess whether her $50 bill will cover her purchase?

   (A) Add up the prices to find the exact cost of $45.67.
   (B) Don't worry about the cost. The $50 should cover the bill.
   (C) Estimate the cost to the nearest dollar, by rounding, to be about $45.00.
   (D) Put back one of the blouses.

2. Alan has nine nickels and pennies worth $0.25. A quick way to calculate the number of each coin would be to_____.

   (A) create a system of equations.
   (B) create a system of inequalities.
   (C) graph two equations, one for pennies and one for nickels.
   (D) use a guess-and-check table.

3. Marco left his sneakers at school, but he cannot remember exactly where in the school they might be. He remembers taking them out of his backpack and putting them in his gym locker. He knows he then took them out of the locker because he used them when he played basketball. He returns to school figuring, correctly, that he left them on the gym floor.

   Marco used what type of reasoning to solve his problem?

   (A) Deductive
   (B) Inductive
   (C) Reductive
   (D) Conductive

4. Good sailing conditions exist when the waves are less than 2 feet high. For today's forecast, the waves are predicted to be 3.5 feet. Given the prediction, most sailors decided to stay home today. The sailors used what type of reasoning to conclude that conditions were poor for sailing?

   (A) Deductive
   (B) Inductive
   (C) Reductive
   (D) Conductive

5. What is the missing reason in the short two-column proof shown below?

Statement	*Reason*
1. $5x + 1 = 36$	Given
2. $5x = 35$	Subtraction Property
3. $x = 7$	?

(A) Division Property
(B) Multiplication Property
(C) Addition Property
(D) Subtraction Property

For questions 6, 7 and 8, use the following data: 12, 5, 37, 48, 9, 3

6. What is the mean of the numbers above?

(A) 12
(B) 19
(C) 37
(D) 48

7. What is the mode of the numbers above?

(A) 5
(B) 12
(C) 48
(D) None

8. What is the median of the numbers above?

(A) 5
(B) 9
(C) 10.5
(D) 12

9. Given the data 12, 14, 18, 18, 19, and 231, which measure would be the least representative of these numbers?

(A) Mean
(B) Mode
(C) Median
(D) Product

10. A mayor of a large city is deciding whether to build a new library, hospital, or city hall. Which one of the following considerations would be most useful in making his decision?

   (A)   His personal desire to build a city hall.

   (B)   An editorial in the newspaper suggesting he decide on a hospital.

   (C)   A close advisor's advice.

   (D)   A survey of the citizens asking them to prioritize the merits of each project.

Answer Key

1. **(C)** By rounding the purchase prices to the nearest dollar, we arrive at the following:

   $12.49 rounds to $12.00

   $13.29 rounds to $13.00

   $19.89 rounds to $20.00

   Notice how close the $45.00 estimate is to the actual price of $45.67.

2. **(D)** Although we could use a system of equations to solve this problem, the small number of coins makes this problem ideal to solve by using a guess-and-check table. Just a few intelligent guesses would yield that four nickels and five pennies will add up to $0.25.

3. **(A)** Marco used deductive reasoning. With deductive reasoning, we use logical principles to draw conclusions. Marco eliminated all the possibilities for the location of the sneakers to arrive at the right conclusion, namely, that he left the sneakers on the gym floor.

4. **(B)** The sailors used inductive reasoning to make their decision not to sail. Inductive reasoning depends upon past observations to make logical decisions. Since past observations suggest that 2-foot waves were the maximum for calm sailing, a predicted 3.5-foot wave would make conditions unsatisfactory.

5. **(A)** To proceed from $5x = 35$ to $x = 7$, we must divide both sides by 5.

   $\dfrac{5x}{5} = \dfrac{35}{5}$ is equivalent to $x = 7$.

6. **(B)** To find the mean of this set of numbers, find their sum and divide by 6.

   Then $\dfrac{12+5+37+48+9+3}{6} = \dfrac{114}{6} = 19.$

7. **(D)** In the set of numbers above, no number occurs more than once. Since no number occurs the most frequently, there is no mode.

8. (C) The median is the number in the middle. First arrange the numbers from least to greatest, so they appear as 3, 5, 9, 12, 37, 48.

Since there are two numbers in the middle, 9 and 12, the median is the mean of these two numbers. Thus, the median is $\dfrac{9+12}{2} = \dfrac{21}{2} = 10.5$.

9. (A) The mean would provide least effective measure of the numbers. The mean is $\dfrac{12+14+18+18+19+231}{6} = \dfrac{312}{6} = 52$.

The mean, 52, is larger than all the elements except for 231. The number 231 skews the data to the larger and is called an outlier.

10. (D) Although each of the selections has some merit, a well-constructed survey of the city's citizens is the best choice. Since the mayor serves the citizens, a well-constructed survey of the citizens' input should be the most influential factor affecting the mayor's decision.

PRACTICE TEST 1

Answer Sheet for Practice Test 1

1. Ⓐ Ⓑ Ⓒ Ⓓ
2. Ⓐ Ⓑ Ⓒ Ⓓ
3. Ⓐ Ⓑ Ⓒ Ⓓ
4. Ⓐ Ⓑ Ⓒ Ⓓ
5. Ⓐ Ⓑ Ⓒ Ⓓ
6. Ⓐ Ⓑ Ⓒ Ⓓ
7. Ⓐ Ⓑ Ⓒ Ⓓ
8. Ⓐ Ⓑ Ⓒ Ⓓ
9. Ⓐ Ⓑ Ⓒ Ⓓ
10. Ⓐ Ⓑ Ⓒ Ⓓ
11. Ⓐ Ⓑ Ⓒ Ⓓ
12. Ⓐ Ⓑ Ⓒ Ⓓ
13. Ⓐ Ⓑ Ⓒ Ⓓ
14. Ⓐ Ⓑ Ⓒ Ⓓ
15. Ⓐ Ⓑ Ⓒ Ⓓ
16. Ⓐ Ⓑ Ⓒ Ⓓ
17. Ⓐ Ⓑ Ⓒ Ⓓ
18. Ⓐ Ⓑ Ⓒ Ⓓ
19. Ⓐ Ⓑ Ⓒ Ⓓ

20. Ⓐ Ⓑ Ⓒ Ⓓ
21. Ⓐ Ⓑ Ⓒ Ⓓ
22. Ⓐ Ⓑ Ⓒ Ⓓ
23. Ⓐ Ⓑ Ⓒ Ⓓ
24. Ⓐ Ⓑ Ⓒ Ⓓ
25. Ⓐ Ⓑ Ⓒ Ⓓ
26. Ⓐ Ⓑ Ⓒ Ⓓ
27. Ⓐ Ⓑ Ⓒ Ⓓ
28. Ⓐ Ⓑ Ⓒ Ⓓ
29. Ⓐ Ⓑ Ⓒ Ⓓ
30. Ⓐ Ⓑ Ⓒ Ⓓ
31. Ⓐ Ⓑ Ⓒ Ⓓ
32. Ⓐ Ⓑ Ⓒ Ⓓ
33. Ⓐ Ⓑ Ⓒ Ⓓ
34. Ⓐ Ⓑ Ⓒ Ⓓ
35. Ⓐ Ⓑ Ⓒ Ⓓ
36. Ⓐ Ⓑ Ⓒ Ⓓ
37. Ⓐ Ⓑ Ⓒ Ⓓ
38. Ⓐ Ⓑ Ⓒ Ⓓ

39. Ⓐ Ⓑ Ⓒ Ⓓ
40. Ⓐ Ⓑ Ⓒ Ⓓ
41. Ⓐ Ⓑ Ⓒ Ⓓ
42. Ⓐ Ⓑ Ⓒ Ⓓ
43. Ⓐ Ⓑ Ⓒ Ⓓ
44. Ⓐ Ⓑ Ⓒ Ⓓ
45. Ⓐ Ⓑ Ⓒ Ⓓ
46. Ⓐ Ⓑ Ⓒ Ⓓ
47. Ⓐ Ⓑ Ⓒ Ⓓ
48. Ⓐ Ⓑ Ⓒ Ⓓ
49. Ⓐ Ⓑ Ⓒ Ⓓ
50. Ⓐ Ⓑ Ⓒ Ⓓ
51. Ⓐ Ⓑ Ⓒ Ⓓ
52. Ⓐ Ⓑ Ⓒ Ⓓ
53. Ⓐ Ⓑ Ⓒ Ⓓ
54. Ⓐ Ⓑ Ⓒ Ⓓ
55. Ⓐ Ⓑ Ⓒ Ⓓ
56. Ⓐ Ⓑ Ⓒ Ⓓ
57. Ⓐ Ⓑ Ⓒ Ⓓ

58. Ⓐ Ⓑ Ⓒ Ⓓ
59. Ⓐ Ⓑ Ⓒ Ⓓ
60. Ⓐ Ⓑ Ⓒ Ⓓ
61. Ⓐ Ⓑ Ⓒ Ⓓ
62. Ⓐ Ⓑ Ⓒ Ⓓ
63. Ⓐ Ⓑ Ⓒ Ⓓ
64. Ⓐ Ⓑ Ⓒ Ⓓ
65. Ⓐ Ⓑ Ⓒ Ⓓ
66. Ⓐ Ⓑ Ⓒ Ⓓ
67. Ⓐ Ⓑ Ⓒ Ⓓ
68. Ⓐ Ⓑ Ⓒ Ⓓ
69. Ⓐ Ⓑ Ⓒ Ⓓ
70. Ⓐ Ⓑ Ⓒ Ⓓ

Practice Test 1—Georgia CRCT

1. What is the value of $\sqrt{3600}$?

   (A) 1800

   (B) 600

   (C) 60

   (D) 6

2. Which of the following is an irrational number?

   (A) $\sqrt{19}$

   (B) 7.6

   (C) $4.\overline{35}$

   (D) $\dfrac{81}{5}$

3. How is "the product of 7 and n decreased by 12" written algebraically?

   (A) $(7 + n) - 12$

   (B) $7n - 12$

   (C) $\dfrac{n}{7} - 12$

   (D) $\dfrac{7}{n} - 12$

GO ON

4. Three points that do not lie on the same line are called _____.

   (A) interior

   (B) non-collinear

   (C) collinear

   (D) exterior

5. Which of the following graphs does not represent a function?

(A)
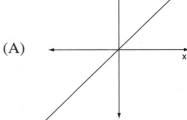

(B)

(C)

(D)

6. Which of the following is a subset of {4, 5, 6, 7, 8}?

    (A) {4, 5, 6, 7, 8}

    (B) {2, 3, 4}

    (C) {0, 1}

    (D) {8, 9, 10}

7. Starting from the point (0, 2), and using the slope of $\frac{2}{3}$, what are the coordinates of the next point that contains integers as x increases?

    (A) (3, 5)

    (B) (2, 5)

    (C) (–2, 0)

    (D) (3, 4)

8. Look at the following diagram of two parallel lines.

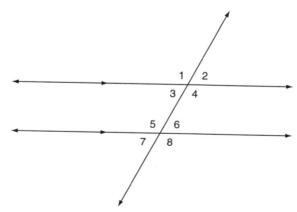

    ∠3 and ∠6 are called _____ angles.

    (A) corresponding

    (B) vertical

    (C) alternate exterior

    (D) alternate interior

GO ON

9. What is the missing step in the two-column proof below?

   Given: $\dfrac{n}{6} = 13$

   Prove: $n = 78$

Statements	Reasons
(1) $\dfrac{n}{6} = 13$	Given
(2) $n = 78$	?

   (A)   Division Property

   (B)   Subtraction Property

   (C)   Multiplication Property

   (D)   Addition Property

10. What is the name of a linear function in the form of $ax + by = c$?

    (A)   Standard Form

    (B)   Slope-Intercept

    (C)   Perpendicular

    (D)   Parallel

11. What is the domain for the function $\{(7, 2)\ (5, 3)\ (8, 2)\ (-4, -4)\}$?

    (A) $\{2, 3, -4\}$

    (B) $\{7, 2\}$

    (C) $\{7, 5, 8, -4\}$

    (D) $\{2, 5, -4\}$

12. What is the length of the missing side in this triangle?

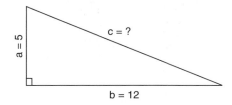

(A) 7

(B) 12

(C) 13

(D) 17

13. What is the value of $\dfrac{(3^2)^2}{3^3}$?

(A) $\dfrac{4}{3}$

(B) 3

(C) 9

(D) 81

14. What is the value of x in the equation $\dfrac{3}{2}x + 2 = 8$?

(A) 4

(B) 6

(C) 8

(D) 9

GO ON

15. Wilma has 3 pairs of red socks, 6 pairs of white socks and 5 pairs of blue socks in her dresser drawer. If she randomly selects one pair of socks, what is the probability that she will select a white pair?

    (A) $\dfrac{1}{7}$

    (B) $\dfrac{3}{7}$

    (C) $\dfrac{3}{5}$

    (D) $\dfrac{6}{5}$

16. For the graph of the equation $y = \dfrac{-2}{3}x + 6,$ what is the y-intercept?

    (A) 6

    (B) $5\dfrac{2}{3}$

    (C) $-\dfrac{2}{3}$

    (D) There is no y-intercept

17. What is (are) the value(s) of n in the equation $|n| + 5 = 12$

    (A) 7

    (B) –7

    (C) Both 7 and –7

    (D) 12

GO ON

18. What is (are) the mode(s) of the following group of numbers?

   7,9,12,12,14,15,15

   (A) No mode
   (B) 12
   (C) 15
   (D) Both 12 and 15

19. $\triangle ABC \cong \triangle DEF$. If $m \angle A = 70°$ and $m\angle D = (4x)°$, what is the value of x?

   (A) 35
   (B) 17.5
   (C) 14
   (D) 7

20. The value of $\sqrt{52}$ lies between which two integers?

   (A) 7 and 8
   (B) 6 and 7
   (C) 5 and 6
   (D) 4 and 5

21. What is the equation of the graph below in slope-intercept form?

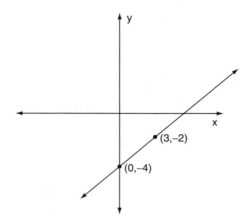

(A) $y = 5x + 4$

(B) $y = \dfrac{5}{2}x + 4$

(C) $y = \dfrac{2}{3}x + 4$

(D) $y = \dfrac{2}{3}x - 4$

22. A local Girl Scout troop uses the model $y = 10x - 5$, where x is the number of cases of cookies sold and y represents the money earned by the troop. If the troop sold 45 cases of cookies, how much money did it earn?

(A) $450

(B) $445

(C) $45

(D) $35

GO ON

23. Return to question #22. If this Girl Scout troop earned $215 in the summer, how many cases did it sell?

    (A) 22

    (B) 21

    (C) 11

    (D) 10

24. A line has how many dimensions?

    (A) 4

    (B) 3

    (C) 2

    (D) 1

25. How many subsets are there for the set {1, 2}?

    (A) 5

    (B) 4

    (C) 3

    (D) 2

26. What type of angles are ∠1 and ∠8 in the following diagram?

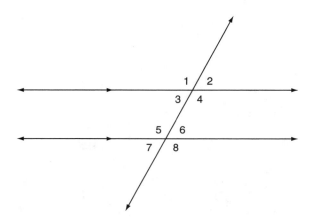

   (A) Alternate exterior

   (B) Alternate consecutive

   (C) Vertical

   (D) Corresponding

27. What is the median of the following group of numbers?

   4, 38, 17, 5, 9, 11

   (A) 14

   (B) 11

   (C) 10

   (D) 9

28. What is the solution to the following system of equations?

$$2x + 4y = 10$$
$$y = 2x - 5$$

(A) $x = 1, y = 3$

(B) $x = 2, y = 2$

(C) $x = 3, y = 1$

(D) $x = 0, y = 2.5$

29. Does the input-output table below express a function?

Input	Output
2	7
−2	−7
3	6
−3	−6

(A) Yes because the inputs and outputs have the same absolute value.

(B) Yes because each input has exactly one output.

(C) No because the table has both negative and positive values.

(D) No because all the outputs are different.

30. If $r \times t = d$, then which of the following is correct?

(A) $r = t \times d$

(B) $r = \dfrac{t}{d}$

(C) $r = t + d$

(D) $r = \dfrac{d}{t}$

31. What is the solution to the inequality $-4m + 12 \geq -20$?

   (A) $m \leq 8$

   (B) $m \geq 8$

   (C) $m \geq 2$

   (D) $m \leq 2$

32. What is the perimeter, in inches, of a square that has an area of 81 square inches?

   (A)   40.5

   (B)   36

   (C)   18

   (D)   9

33. A diagram that shows set relationships is called a _____ diagram.

   (A)   Subset

   (B)   Union

   (C)   Intersection

   (D)   Venn

34. What is the value of $\sqrt{49} + (-\sqrt{16}) + \sqrt{\dfrac{9}{25}}$?

    (A) $3\dfrac{3}{5}$

    (B) $33\dfrac{9}{25}$

    (C) $55\dfrac{3}{25}$

    (D) 63

35. The number $\sqrt{96}$ is equivalent to which of the following?

    (A) 48

    (B) $3\sqrt{32}$

    (C) $4\sqrt{6}$

    (D) $8\sqrt{3}$

36. Geometric figures with corresponding side lengths and angles of equal measures are called _____.

    (A) obtuse

    (B) acute

    (C) parallel

    (D) congruent

GO ON

37. Theresa reviewed the following input-output table relating gasoline prices and her mother's recent purchases.

Input	Output
Gas Price (per gallon)	Gallons Purchased
$3.81	12
$4.01	9
$3.26	14
$2.89	18

What conclusion can Theresa draw regarding her mother's gasoline purchases and gasoline's price?

(A)   The price of gasoline will always increase.

(B)   The cheaper the cost of gasoline, the more gallons her mother would purchase.

(C)   The more expensive the cost of gasoline, the more gallons her mother purchased.

(D)   Her mother should never purchase more than 10 gallons of gasoline.

38. When shopping, the most sensible way to estimate your cost at the register is to

(A)   let the cashier ring up the cost and see if you have the right amount of money.

(B)   start putting items away if you think you don't have enough money.

(C)   round off the cost of each item to the nearest dollar.

(D)   ask the next person in line if he/she has a calculator.

GO ON

39. In standard form, the equation $y = \dfrac{-2}{3}x + 6$ becomes _____.

(A) $\dfrac{2}{3}x + y = 6$

(B) $2x + 3y = 18$

(C) $\dfrac{3}{2}x + y = 12$

(D) $3x + 2y = 12$

40. If the points $(-1, -3)$, $(0, -1)$, $(1, 1)$, and $(2, 3)$ are graphed, what type of function is shown?

(A) Quadratic

(B) Linear

(C) Vertical

(D) Transverse

41. If $X = \{4, 5, 6, 7\}$ and $Y = \{6, 7, 8, 9)$, which of the following represents $X \cup Y$?

(A) $\{6, 7\}$

(B) $\{4, 9\}$

(C) $\{4, 5, 6, 7, 8, 9\}$

(D) $\{4, 5, 8, 9\}$

GO ON

42. Look at the following diagram of two parallel lines.

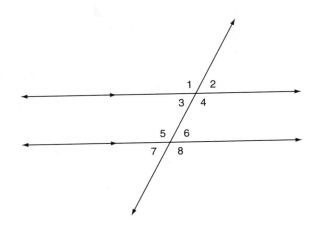

If $m \angle 3 = 48°$ and the measure of $m \angle 6 = (6x)°$, what is the value of x?

(A)   8

(B)   12

(C)   16

(D)   24

43. The relation $\{(2, 5) (3, 5) (8, 5) (9, 5)\}$ can be described as_____.

   (A)   not a function because the output is repeated.

   (B)   not a function because the difference between inputs is different.

   (C)   a function because the inputs are in ascending order.

   (D)   a function because each input has exactly one output.

44. A salesperson gets a $12 commission on all sales, plus a $100 per week salary. If she made 30 sales in a week, what would be her earnings?

    (A) $360

    (B) $420

    (C) $460

    (D) $480

45. What is the inequality that corresponds to the graph shown below?

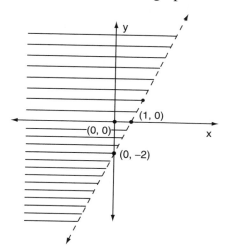

    (A) $y < 2x - 2$

    (B) $y > 2x - 2$

    (C) $y \geq 2x - 2$

    (D) $y \leq 2x - 2$

GO ON

46. In which order would the numbers $-7.1, \sqrt{51}, 7$, and 0.7 appear, when arranged from least to greatest?

   (A) $-7.1, \sqrt{51}, 7, 0.7$
   (B) $-7.1, 0.7, \sqrt{51}, 7$
   (C) $-7.1, 0.7, 7, \sqrt{51}$
   (D) $-7.1, 7, 0.7, \sqrt{51}$

47. A coin is flipped four times. What is the probability that four tails will appear?

   (A) $\dfrac{1}{16}$
   (B) $\dfrac{1}{8}$
   (C) $\dfrac{1}{4}$
   (D) $\dfrac{1}{2}$

48. What is an equivalent expression for $\sqrt{48} + 3\sqrt{3}$?

   (A) $\sqrt{54}$
   (B) $3\sqrt{51}$
   (C) $7\sqrt{3}$
   (D) $3\sqrt{7}$

GO ON

49. What is the mean of the numbers 18, 5, 34, 87, and 99?

(A) 32

(B) 34

(C) 42.4

(D) 48.6

50. Which of the following statements is true concerning the intersections of perpendicular lines?

(A) They never intersect.

(B) They form right angles.

(C) They form only obtuse angles.

(D) They intersect at two points.

51. If $R = \{5, 6, 7, 8, 9\}$ and $S = \{2, 7, 9, 10, 11\}$, what is the representation of $R \cap S$?

(A) $\{7, 9\}$

(B) $\{2, 5, 6, 7, 8, 9, 10, 11\}$

(C) $\varnothing$

(D) $\{5, 6\}$

GO ON

52. Which inequality describes the graph below?

(A) $x \geq 6$

(B) $x \leq 6$

(C) $x < 6$

(D) $x > 6$

53. What is the solution to the following system of equations?

$4x + y = 7$

$2x - y = 5$

(A) $x = 1, y = -2$

(B) $x = -1, y = 2$

(C) $x = 2, y = -1$

(D) $x = 2, y = 2$

54. What is the slope of the line corresponding to the equation $y = \dfrac{-5}{2}x + 6$?

(A) $\dfrac{-5}{2}$

(B) $-\dfrac{2}{5}$

(C) $\dfrac{7}{2}$

(D) 6

55. What is the range of the relation {(5, 2), (5, 3), (–2, –4), (4, 4)}?

(A) {5, –2, 4}

(B) {2, 3, –4, 4}

(C) {5, –2}

(D) {5, 2, 3}

56. What is the domain of the relation of the relation {(6, 2) (5, 3) (7, 5) (3, 1)}?

(A) {5, –2, 4}

(B) {6, 5, 7, 3}

(C) {5}

(D) {–4, 4}

57. Start from the point (4, –6), and use a slope of 2. As x increases, what would be the next point that contains integer values for x and y?

(A) (6, –5)

(B) (6, –6)

(C) (5, –4)

(D) (4, –5)

58. The _____ line test determines if a graph is a function.

(A) diagonal

(B) parallel

(C) horizontal

(D) vertical

GO ON

59. If $A = \{5, 6, 7, 8\}$ and $B = \{6, 7\}$, which statement illustrates the relationship between A and B?

   (A) $A \in B$
   (B) $A \subseteq B$
   (C) $B \subseteq A$
   (D) $A = B$

60. A conclusion based on a survey is an example of which type of reasoning?

   (A) Productive
   (B) Conductive
   (C) Deductive
   (D) Inductive

61. What is the scientific notation for the number 0.0086?

   (A) 8.6×10^{-2}
   (B) 86×10^{-4}
   (C) 8.6×10^{-3}
   (D) 86×10^{-5}

62. An irrational number has a _____.

   (A) pattern that does not repeat
   (B) pattern that does repeat
   (C) terminating decimal
   (D) fractional equivalent

GO ON

63. Teddy has fewer than 20 coins in his pocket. Each coin is either a nickel or a dime. The total value of these coins exceeds $1.25. Let x represent the number of nickels and let y represent the number of dimes. Which of the following pairs of inequalities would represent the given information?

(A) $x + y > 20$
$0.05x + 0.10y < 1.25$

(B) $x + y > 20$
$0.10x + 0.05y < 1.25$

(C) $x + y < 20$
$0.10x + 0.05y > 1.25$

(D) $x + y < 20$
$0.05x + 0.10y > 1.25$

64. What is the simplified form for $4\sqrt{5} - 7\sqrt{5} - \sqrt{5}$?

(A) $-3\sqrt{5}$

(B) $-4\sqrt{5}$

(C) $-3\sqrt{15}$

(D) $-4\sqrt{15}$

65. Consider the following four inequalities:

I: $-3x + 4 < 6$
II: $-3x + 4 < -6$
III: $3x - 4 < 6$
IV: $3x - 4 < -6$

For how many of these inequalities would the solution set be represented by x greater than a number?

(A) 4

(B) 3

(C) 2

(D) 1

66. What is the value of y in the following system of equations?

$2x - y = 9$

$y = 4x - 15$

    (A) 4

    (B) 3

    (C) −3

    (D) −4

67. If $b = c - 2f$, which of the following expressions is equivalent to f?

    (A) $\dfrac{b-c}{-2}$

    (B) $\dfrac{b+c}{-2}$

    (C) $\dfrac{b}{c-2}$

    (D) $\dfrac{b}{c+2}$

68. What is the simplified form for $\dfrac{60\sqrt{50}}{30\sqrt{2}}$?

    (A) 150

    (B) 10

    (C) $120\sqrt{3}$

    (D) $8\sqrt{3}$

GO ON

69. The Long Ears phone company has the following rates for long distance calls:

A flat rate of $3.00 for the first 30 minutes.

$0.06 per minute for each additional minute.

Let x represent the number of minutes of a phone call and let y represent the total cost. Assuming that $x > 30$, which equation describes the total cost in terms of the number of minutes?

(A) $y = (30)(3) + 0.06x$

(B) $y = (30)(3 + 0.06)x$

(C) $y = 30 + 0.06(x{-}3)$

(D) $y = 3 + 0.06\,(x - 30)$

70. The relation $\{(6, 4), (5, 1), \text{and } (x, 7)\}$ is *not* a function. Which of the following could represent the value of x?

(A)   1

(B)   4

(C)   6

(D)   7

Answer Key—Practice Test 1

Item Number	Correct Answer	Domain Description	Standard Measured
01	C	Numbers and Operations	M8N1
02	A	Numbers and Operations	M8N1
03	B	Algebra	M8A1
04	B	Geometry	M8G1
05	C	Algebra	M8A3
06	A	Data Analysis and Probability	M8D1
07	D	Algebra	M8A4
08	D	Geometry	M8G1
09	C	Algebra	M8P2
10	A	Algebra	M8A4
11	C	Algebra	M8A3
12	C	Geometry	M8G2
13	B	Numbers and Operations	M8N1
14	A	Algebra	M8A1
15	B	Data Analysis and Probability	M8D3
16	A	Algebra	M8A4
17	C	Algebra	M8A1
18	D	Data Analysis and Probability	M8P5
19	B	Geometry	M8G1
20	A	Numbers and Operations	M8N1
21	D	Algebra	M8A4
22	B	Algebra	M8A1
23	A	Algebra	M8A1
24	D	Geometry	M8G1
25	B	Data Analysis and Probability	M8D1

26	A	Geometry	M8G1
27	C	Data Analysis and Probability	M8P5
28	C	Algebra	M8A5
29	B	Algebra	M8A3
30	D	Algebra	M8A1
31	A	Algebra	M8A2
32	B	Numbers and Operations	M8N1
33	D	Data Analysis and Probability	M8D1
34	A	Numbers and Operations	M8N1
35	C	Numbers and Operations	M8N1
36	D	Geometry	M8G1
37	B	Algebra	M8A3
38	C	Numbers and Operations	M8P1
39	B	Algebra	M8A4
40	B	Algebra	M8A3
41	C	Data Analysis and Probability	M8D1
42	A	Geometry	M8G1
43	D	Algebra	M8A3
44	C	Algebra	M8A1
45	B	Algebra	M8A2
46	C	Numbers and Operations	M8N1
47	A	Data Analysis and Probability	M8D3
48	C	Numbers and Operations	M8N1
49	D	Data Analysis and Probability	M8P5
50	B	Numbers and Operations	M8N1
51	A	Data Analysis and Probability	M8D1
52	B	Algebra	M8A2
53	C	Algebra	M8A5
54	A	Algebra	M8A4
55	B	Algebra	M8A3
56	B	Algebra	M8A3

57	C	Algebra	M8A4
58	D	Algebra	M8A3
59	C	Data Analysis and Probability	M8D1
60	D	Data Analysis and Probability	M8P5
61	C	Numbers and Operations	M8N1
62	A	Numbers and Operations	M8N1
63	D	Algebra	M8A5
64	B	Numbers and Operations	M8N1
65	C	Algebra	M8A2
66	C	Algebra	M8A5
67	A	Algebra	M8A1
68	B	Numbers and Operations	M8N1
69	D	Algebra	M8A4
70	C	Algebra	M8A3

Practice Test 1—Progress Chart— Georgia CRCT

Numbers and Operations   ___/ 15

01	02	13	20	32	34	35	38	46	48

50	61	62	64	68

Geometry   ___/ 8

04	08	12	19	24	26	36	42

Algebra   ___/ 35

03	05	07	09	10	11	14	16	17	21

22	23	28	29	30	31	37	39	40	43

44	45	52	53	54	55	56	57	58	63

65	66	67	69	70

Data Analysis and Probability ___/ 12

05	07	08	14	16	25	26	34	41

51	59	60

Total ___/ 70

Detailed Solutions

1. **(C)** $\sqrt{3600} = 60$, since $(60)(60) = 3600$. 60 is called the square root of 3600.

2. **(A)** The decimal form of an irrational number neither ends nor repeats a pattern. $\sqrt{19}$ is approximately equal to 4.3588989…

3. **(B)** A product is the answer to a multiplication problem. "Decreased by" means subtraction is taking place.

4. **(B)** By definition, three points that do not lie on the same line are called non-collinear points.

5. **(C)** The graph in selection C does not pass the vertical line test, as shown below.

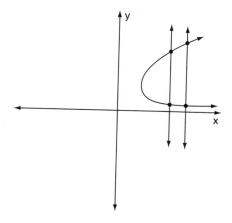

6. **(A)** Every set is a subset of itself.

7. **(D)** Slope can be described as $\dfrac{rise}{run}$. From the point (0, 2), rise 2 and run 3 to arrive at (3, 4). This is illustrated in the figure below.

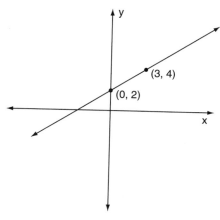

8. **(D)** $\angle 3$ and $\angle 6$ are called alternate interior angles because they are within the parallel lines and on alternate sides of the transversal.

9. **(C)** In order to proceed from $\dfrac{n}{6} = 13$ to $n = 78$, we must have multiplied both sides by 6. Then $(6)\left(\dfrac{n}{6}\right) = (13)(6)$, so $n = 78$.

10. **(A)** By definition, a linear function in the form of $ax + by = c$ is called the standard form of a line. Also, a, b, and c must all be integers and a is usually positive.

11. **(C)** The domain is the set of all the inputs, which are the first coordinates of each ordered pair.

12. **(C)** The missing side, called the hypotenuse, can be found as follows.

$$5^2 + 12^2 = c^2$$
$$25 + 144 = c^2$$
$$169 = c^2$$
$$13 = c$$

13. **(B)** $\dfrac{(3^2)^2}{3^3} = \dfrac{(9)^2}{27} = \dfrac{81}{27} = 3 \cdot$

14. **(A)** These are the steps for finding x.

$$\frac{3}{2}x + 2 = 8$$
$$\frac{3}{2}x + 2 - 2 = 8 - 2$$
$$\frac{3}{2}x = 6$$
$$\frac{2}{3}\left(\frac{3}{2}x\right) = \frac{2}{3}(6)$$
$$x = 4$$

15. **(B)** There are 6 pairs of white socks and a total of 14 pairs of socks. Then the required probability is $\dfrac{6}{14} = \dfrac{3}{7} \cdot$

16. **(A)** In the equation $y = mx + b$, m represents the slope and b represents the y-intercept. Thus, in the equation $y = \dfrac{-2}{3}x + 6$, 6 is b, the y-intercept.

17. (C)   Solve for n by isolating the absolute value.

$$|n| + 5 = 12$$
$$|n| + 5 - 5 = 12 - 5$$
$$|n| = 7$$

The absolute value represents the positive distance a number is from 0, so n could be 7 or -7.

18. (D)   A mode of a group of numbers is the value that occurs most frequently.

Since both 12 and 15 occur twice in the list, they are both modes.

19. (B)   Since $\triangle ABC \cong \triangle DEF$, the $m\angle A = m\angle D$. Therefore $70 = 4x$. Divide both sides by 4, so that $x = 17.5$.

20. (A)   Since $\sqrt{49} < \sqrt{52} < \sqrt{64}$, we conclude that $7 < \sqrt{52} < 8$.

21. (D)   The line intersects the y-axis at $(0, -4)$ so the y-intercept is b. From $(0, -4)$, the rise is 2 units and the run is 3 units in order to arrive at $(3, -2)$. Thus, the slope is $\dfrac{2}{3}$. Inserting the slope and the y-intercept values into the formula, the equation for the graph is $y = \dfrac{2}{3}x - 4$.

22. (B)   Substitute 45 for x in the equation.

$$y = 10(45) - 5$$
$$y = 445$$

23. (A)   Substitute 215 for y in the equation.

$$215 = 10x - 5$$
$$215 + 5 = 10x - 5 + 5$$
$$220 = 10x$$
$$\frac{220}{10} = \frac{10x}{10}$$
$$x = 22$$

24. (D)   A line has one dimension, namely length.

25. (B)   The subsets are $\{1\}, \{2\}, \{1, 2\}$ and $\varnothing$.

26. (A)   By definition, $\angle 1$ and $\angle 8$ are alternate exterior angles.

27. (C)   Arrange the numbers from least to greatest, so that the list of numbers is 4, 5, 9, 11, 17, 38. Then the two middle numbers are 9 and 11. Thus, the median is $\dfrac{9 + 11}{2} = 10$.

28. (C)   Substitute $2x - 5$ for y in the first equation.

$$2x + 4(2x - 5) = 10$$
$$2x + 8x - 20 = 10$$
$$10x - 20 = 10$$
$$10x - 20 + 20 = 10 + 20$$
$$10x = 30$$
$$\frac{10x}{10} = \frac{30}{10}$$
$$x = 3$$

Now, substitute 3 for x in either equation. We'll choose the second equation.

$$2(3) + 4y = 10$$
$$6 + 4y = 10$$
$$6 - 6 + 4y = 10 - 6$$
$$4y = 4$$
$$\frac{4y}{4} = \frac{4}{4}$$
$$y = 1$$

29. (B)   The input-output table expresses a function. A function is a relation in which each input has exactly one output.

30. (D)   Here are the steps in solving for r.

$$r \times t = d$$
$$\frac{r \times t}{t} = \frac{d}{t}$$
$$r = \frac{d}{t}$$

31. (A)   Here are the steps.

$$-4m + 12 \geq -20$$
$$-4m + 12 - 12 \geq -20 - 12$$
$$-4m \geq -32$$
$$\frac{-4m}{-4} \leq \frac{-32}{-4}$$
$$m \leq 8$$

Remember to reverse the direction of the inequality when dividing or multiplying by a negative number.

32. (B) Calculate the square root of 81 to find the length of one side of the square, which is $\sqrt{81} = 9$. Since a square has four sides of equal length, multiply 9 by $4 = 36$ to find the perimeter of the square.

33. (D) By definition, a Venn diagram shows set relationships.

34. (A) $\sqrt{49} + (-\sqrt{16}) + \sqrt{\dfrac{9}{25}} = 7 - 4 + \dfrac{3}{5} = 3\dfrac{3}{5}.$

35. (C) $\sqrt{96} = \sqrt{16} \times \sqrt{6} = 4\sqrt{6}.$

36. (D) By definition, geometric figures with corresponding sides and angles having equal measures are called congruent figures.

37. (B) As the cost of gasoline became cheaper, Theresa's mother would purchase more gasoline.

38. (C) A sensible way to assess if you have enough money to make your purchases is to estimate the cost of the goods by rounding each item to the nearest dollar. Often, we do not have a paper and pencil or a calculator when shopping.

39. (B) The standard form of a line expresses an equation in the form of $ax + by = c$, where a, b and c are integers. Here are the steps required.

$$y = \frac{-2}{3}x + 6$$

$$\frac{2}{3}x + y = \frac{-2}{3}x + \frac{2}{3}x + 6$$

$$\frac{2}{3}x + y = 6$$

$$3\left(\frac{2}{3}x + y = 6\right)$$

$$2x + 3y = 18$$

40. (B) When graphed, the points lie along a straight line, as shown below.

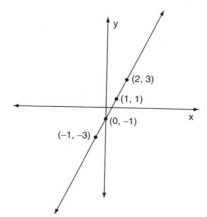

Therefore the function is linear.

41. (C) $X \cup Y = X + Y - X \cap Y = \{4, 5, 6, 7\} + \{6, 7, 8, 9\} - \{6, 7\} = \{4, 5, 6, 7, 8, 9\}$. This is equivalent to stating that $X \cup Y$ consists of all elements that belong to X, to Y, or to both X and Y.

42. (A) $\angle 3$ and $\angle 6$ are alternate interior angles. Their measures are equal when the lines are parallel. Thus, $6x = 48$, so $x = \dfrac{48}{6} = 8$.

43. (D)   The relation is a function because each input has exactly one output. A relation can be a function if the outputs are repeated but the inputs cannot repeat.

44. (C)   Multiply 30 by 12 and add 100. Then $(12)(30) + 100 = 460$. Note that this situation could be modeled as the linear function $y = 12x + 100$, for which x represents the number of sales and y represents the total earnings.

45. (B)   Since the line in the inequality is dotted, both answers C and D are eliminated. A dotted line means that a graph can be greater than or less than, but not equal to the values contained on the line. Now test selection B using the point $(0, 0)$, which means that we check if $0 > (2)(0) - 2$. This inequality simplifies to $0 > -2$, which is true.

46. (C)   Since $\sqrt{49} = 7$, we know that $\sqrt{51} > 7$. Thus $\sqrt{51}$ is the largest of the four values. The only negative number is -7.1, so it must be the smallest of these numbers. Finally, we note that $0.7 < 7$.

47. (A)   Each coin flip is an independent event; one flip has no effect on the next flip. Since each flip has a probability of $\dfrac{1}{2}$ of landing on heads, multiply $\dfrac{1}{2}$ times itself 4 times. Thus, the required probability is $\dfrac{1}{2} \times \dfrac{1}{2} \times \dfrac{1}{2} \times \dfrac{1}{2} = \dfrac{1}{16}$.

48. (C) $\sqrt{48} + 3\sqrt{3} = \sqrt{16} \times \sqrt{3} + 3\sqrt{3} = 4\sqrt{3} + 3\sqrt{3} = 7\sqrt{3}$.

49. (D)   The mean is calculated as $\dfrac{18 + 5 + 34 + 87 + 99}{5} = \dfrac{243}{5} = 48.6$.

50. (B)   By definition, perpendicular lines intersect to form right angles. In the following diagram, $\overleftrightarrow{AB} \perp \overleftrightarrow{CD}$.

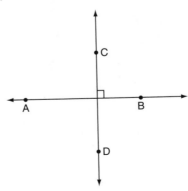

51. (A) The symbol "∩" means "intersect." The intersection of the two sets consists of the elements common to both sets. In this example, {7, 9} represents the elements common to R and S.

52. (B) The graph indicates all values smaller than 6 are part of the solution. The solid dot at 6 means that 6 is also part of the solution.

53. (C) Add the two equations downward to eliminate y.

$$4x + y = 7$$
$$+2x - y = 5$$
$$6x = 12$$
$$x = 2$$

Now, substitute $x = 2$ in either equation. We'll choose the first equation.

Then, $4(2) + y = 7$, which simplifies to $8 + y = 7$. Thus, $y = 7 - 8 = -1$.

54. (A) In the form $y = mx + b$, the slope is the value of m.

55. (B) In any relation, the range is the collection of outputs, which is represented by the set of y values of each ordered pair. Thus, the range is {2, 3, –4, 4}.

56. (B) The domain in any relation is the collection of inputs, which is represented by the set of x values of each ordered pair. Thus, the domain is {6, 5, 7, 3}.

57. (C) A slope of 2 means a rise of 2 and a run of 1, as shown by the following graph.

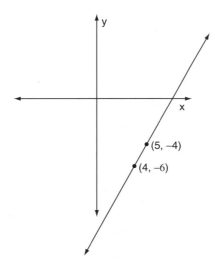

Thus, the next point is (5, –4).

58. (D) If the intersection of any vertical line and a graph never occurs or occurs only once, then the graph must represent a function.

59. (C) All the elements of B are contained in A, so B is a subset of A. The symbol "⊆" means "is a subset of."

60. (D) The results of a survey indicate what a population at large may think. Drawing a conclusion based on specific cases, such as survey results, is an example of inductive reasoning.

61. (C) In order to write the number in scientific notation, first place the decimal point between 8 and 6 so that the number part is between 1 and 10. To go from 8.6 to 0.0086, the decimal point must be moved three places to the left. This means that 8.6 is being multiplied by 0.001, which is 10^{-3}. Thus, the answer is 8.6×10^{-3}.

62. (A) An irrational number, when written as a decimal, shows a pattern that neither terminates nor repeats.

63. (D) The symbol $<$ is used for the phrase "fewer than," so $x + y < 20$. The value x nickels is represented by $0.05x$. The value of y dimes is represented by $0.10y$. The symbol $>$ is used for the word "exceeds." Thus, the second inequality becomes $0.05x + 0.10y > 1.25$.

64. (B) When adding or subtracting like terms involving square roots, just add or subtract the coefficients. Thus, remembering that $-\sqrt{5}$ is equivalent to $-1\sqrt{5}$, the answer becomes $(4-7-1)(\sqrt{5}) = -4\sqrt{5}$.

65. (C) For inequalities I and II, the last step in solving for x involves a division by –3. Whenever an inequality involves the operation of either multiplication or division by a negative number, the order of inequality must be reversed. The answers for I and II are $x > -\dfrac{2}{3}$ and $x > \dfrac{10}{3}$, respectively.

66. (C) Rewrite the second equation as $-4x + y = -15$. Add this equation to the first equation $2x - y = 9$ to get $-2x = -6$. Then $x = \dfrac{-6}{-2} = 3$. Substituting this value of x into the second equation yields the value of $y = (4)(3) - 15 = -3$.

67. (A) Subtract c from each side of the equation. Then $b - c = -2f$. The next step is to divide both sides by the coefficient of f, which is –2. Thus, $\dfrac{b-c}{-2} = f$.

68. (B) $\dfrac{60}{30} = 2$ and $\dfrac{\sqrt{50}}{\sqrt{2}} = \sqrt{25} = 5$. Then the expression is equal to $(2)(5) = 10$.

69. (D) In order to find the total cost, y, we first recognize that $3.00 is the cost for the first 30 minutes. Then, we calculate the number of minutes used beyond 30 minutes. If x represents the total number of minutes, then each of these additional minutes costs 0.06. Thus, the total cost is given by the expression $3 + (0.06)(x - 30)$.

70. (C) A relation is not a function if at least two of its ordered pairs contain the same domain value but a different range value. If $x = 6$, then both (6, 4) and (6, 7) belong to this relation. The same domain value of 6 would have two range values.

PRACTICE TEST 2

Answer Sheet for Practice Test 2

1. (A) (B) (C) (D)   20. (A) (B) (C) (D)   39. (A) (B) (C) (D)   58. (A) (B) (C) (D)

2. (A) (B) (C) (D)   21. (A) (B) (C) (D)   40. (A) (B) (C) (D)   59. (A) (B) (C) (D)

3. (A) (B) (C) (D)   22. (A) (B) (C) (D)   41. (A) (B) (C) (D)   60. (A) (B) (C) (D)

4. (A) (B) (C) (D)   23. (A) (B) (C) (D)   42. (A) (B) (C) (D)   61. (A) (B) (C) (D)

5. (A) (B) (C) (D)   24. (A) (B) (C) (D)   43. (A) (B) (C) (D)   62. (A) (B) (C) (D)

6. (A) (B) (C) (D)   25. (A) (B) (C) (D)   44. (A) (B) (C) (D)   63. (A) (B) (C) (D)

7. (A) (B) (C) (D)   26. (A) (B) (C) (D)   45. (A) (B) (C) (D)   64. (A) (B) (C) (D)

8. (A) (B) (C) (D)   27. (A) (B) (C) (D)   46. (A) (B) (C) (D)   65. (A) (B) (C) (D)

9. (A) (B) (C) (D)   28. (A) (B) (C) (D)   47. (A) (B) (C) (D)   66. (A) (B) (C) (D)

10. (A) (B) (C) (D)   29. (A) (B) (C) (D)   48. (A) (B) (C) (D)   67. (A) (B) (C) (D)

11. (A) (B) (C) (D)   30. (A) (B) (C) (D)   49. (A) (B) (C) (D)   68. (A) (B) (C) (D)

12. (A) (B) (C) (D)   31. (A) (B) (C) (D)   50. (A) (B) (C) (D)   69. (A) (B) (C) (D)

13. (A) (B) (C) (D)   32. (A) (B) (C) (D)   51. (A) (B) (C) (D)   70. (A) (B) (C) (D)

14. (A) (B) (C) (D)   33. (A) (B) (C) (D)   52. (A) (B) (C) (D)

15. (A) (B) (C) (D)   34. (A) (B) (C) (D)   53. (A) (B) (C) (D)

16. (A) (B) (C) (D)   35. (A) (B) (C) (D)   54. (A) (B) (C) (D)

17. (A) (B) (C) (D)   36. (A) (B) (C) (D)   55. (A) (B) (C) (D)

18. (A) (B) (C) (D)   37. (A) (B) (C) (D)   56. (A) (B) (C) (D)

19. (A) (B) (C) (D)   38. (A) (B) (C) (D)   57. (A) (B) (C) (D)

Practice Test 2—Georgia CRCT

1. What is the simplified form for the value of $\sqrt{80} \times \sqrt{5}$?

   (A) $\sqrt{85}$

   (B) $4\sqrt{100}$

   (C) 200

   (D) 20

2. Which of the following statements is true concerning parallel lines?

   (A) They never intersect.

   (B) They intersect to form complementary angles.

   (C) They intersect to form right angles.

   (D) They never have a slope.

3. Which equation does the following statement describe?

   "The difference of a number n and twelve is six."

   (A) $12 - n = 6$

   (B) $n - 12 = 6$

   (C) $n - 6 = 12$

   (D) $12 - 6 = n$

GO ON

4. Does the following graph express a function?

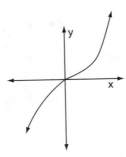

(A) Yes, because it is curved.

(B) No, because it is curved.

(C) Yes, because it passes the vertical line test.

(D) No, because it fails the vertical line test.

5. A lunch special offers three different sandwiches, three different side orders, and five different beverages. How many different lunch combinations are available to each diner?

(A) 11

(B) 12

(C) 45

(D) 54

6. What is the solution to the following system of equations?

$y = 2x - 1$

$2x + 3y = 21$

(A) $x = 3$ and $y = 5$

(B) $x = 5$ and $y = 3$

(C) $x = 7$ and $y = 2$

(D) $x = 2$ and $y = 7$

GO ON

7. A line that most closely passes through the points of a scatter plot is called a _____.

    (A) line of best fit
    (B) parallel line
    (C) perpendicular line
    (D) transversal

8. What is the mode for the following group of numbers?

    7,8,8,9,9,9,10

    (A) 7
    (B) 8
    (C) 9
    (D) 10

9. How is the number 1,234,000 written in scientific notation?

    (A) 12.34×10^5
    (B) 123.4×10^4
    (C) 0.1234×10^7
    (D) 1.234×10^6

GO ON

10. In the figure below, $m \angle 1 = 80°$.

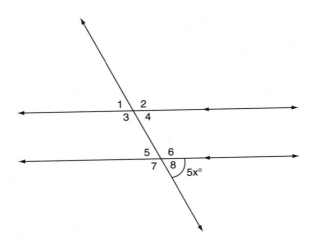

What is the value of x?

(A) 12

(B) 16

(C) 40

(D) 80

11. Given $2br = 7mn$, what is the expression for r in terms of b, m, and n?

(A) $\dfrac{7mn}{2b}$

(B) $\dfrac{2}{7}mnb$

(C) $\dfrac{7b}{2mn}$

(D) $\dfrac{7}{2}mnb$

GO ON

12. Does the following input-output table express a function?

Input	Output
7	6
8	5
7	4
5	3

   (A) Yes, because each output is different.

   (B) Yes, because 7 appears twice as an input.

   (C) No, because if the input increases, the output should increase.

   (D) No, because there is more than one output for the input 7.

13. What is the slope of the graph of the equation $4x + 3y = 12$?

   (A) $\dfrac{4}{3}$

   (B) $\dfrac{-4}{3}$

   (C) $\dfrac{3}{4}$

   (D) $\dfrac{-3}{4}$

GO ON

14. Consider the Venn diagram shown below.

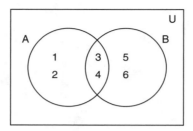

Which set describes $\overline{A}$?

(A) {1, 2}

(B) {3, 4}

(C) {5, 6}

(D) {1, 2, 5, 6}

15. A Boy Scout troop raised funds by running a 10K race. Each boy would receive a $10 donation plus $2 for each mile run. If *x* represents the number of miles and *y* represents the corresponding amount of money, which linear function models this situation?

(A) $y = 2x + 10$

(B) $x = 2y + 10$

(C) $y = 10x + 2$

(D) $x = 10y + 2$

16. What is the median of the following group of numbers?

9, 9, 9, 10, 10, 37

(A) 9

(B) 9.5

(C) 10

(D) 14

17. In the following two-column proof, what is the missing statement?

Given: $7x = 98$

Prove: $x = 14$

Statement	*Reason*
(1) $7x = 98$	(1) Given
(2) ?	(2) Division Property

(A) $x = 12$

(B) $x = 13$

(C) $x = 14$

(D) $x = 2$

18. What is the simplified form for the number $\sqrt{147}$?

(A) 73.5

(B) 12

(C) $3\sqrt{7}$

(D) $7\sqrt{3}$

GO ON

19. What is the missing length in the triangle shown below?

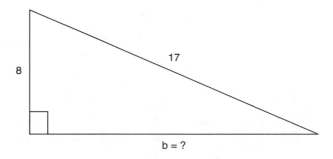

(A) 15

(B) 9

(C) 8

(D) 6

20. What is the value of x in the equation $x - (-52) = -43$?

(A) 9

(B) −95

(C) −9

(D) −7

21. Which of the following graphs describes the inequality x < −2?

(A) ![number line: -3 -2 -1 0 1 2 with filled dot at -2]

(B) ![number line: -3 -2 -1 0 1 2 with filled dot at -2]

(C) ![number line: -2 -1 0 1 2 with open circle at -2]

(D) ![number line: -3 -2 -1 0 1 2 with open circle at -2]

22. What is the slope of the graph shown below?

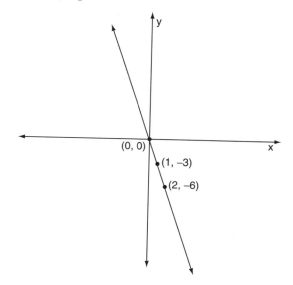

(A) 3

(B) −3

(C) $\dfrac{1}{3}$

(D) $\dfrac{-1}{3}$

23. What is the *y*-intercept in the graph of the equation $y = -5x + 4$?

(A) -5

(B) $\dfrac{-5}{4}$

(C) 4

(D) $\dfrac{-4}{5}$

24. Which inequality describes the graph shown below?

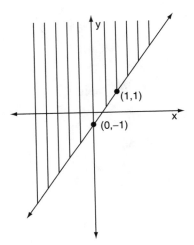

(A) $y > 2x - 1$

(B) $y < 2x - 1$

(C) $y \geq 2x - 1$

(D) $y \leq 2x - 1$

GO ON

25. If $P = \{13,15,17\}$ and $R = \{2,4,6,8\}$, which set describes $P \cap R$?

    (A) $\varnothing$

    (B) $\{2, 4, 6, 8, 13, 15, 17\}$

    (C) $\{13, 2\}$

    (D) $\{17, 8\}$

26. For the following group of numbers, what two measures are the same?

1,3,5,7,9

    (A) Mode and median

    (B) Mean and median

    (C) Mode and mean

    (D) Domain and range

27. Between which two whole numbers does $\sqrt{129}$ lie?

    (A) 12 and 13

    (B) 11 and 12

    (C) 10 and 11

    (D) 9 and 10

28. What is the value of $(-2)^5$?

    (A) 10

    (B) -10

    (C) 32

    (D) -32

GO ON

29. Consider the diagram shown below.

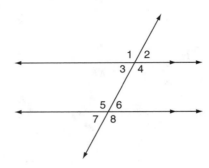

∠4 and ∠8 are called _____ angles.

(A) Corresponding

(B) Alternate interior

(C) Alternate exterior

(D) Vertical

30. Sammy can spend up to $40 on his next pair of sneakers. Which inequality represents this situation?

(A) $x > 40$

(B) $x \leq 40$

(C) $x \geq 40$

(D) $x < 40$

31. The Atlanta Swim Club raises money by selling racing swimsuits. The equation that shows the money raised (y) for the number of swimsuits sold (x) is $y = 18x$. How many swimsuits were sold to raise $774?

(A) 74

(B) 44

(C) 43

(D) 18

GO ON

32. For the graph of the equation $6x + 4y = 12$, what is the x-intercept?

   (A) $(0, 2)$

   (B) $(0, 3)$

   (C) $(2, 0)$

   (D) $(3, 0)$

33. What system of inequalities is illustrated by the graph shown below?

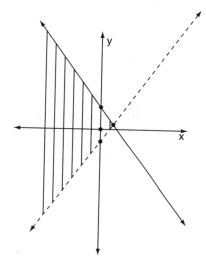

   (A) $y > 2x - 1$
   $y \leq -2x + 3$

   (B) $y \geq 2x - 1$
   $y < -2x + 3$

   (C) $y \leq 2x - 1$
   $y \leq -2x + 3$

   (D) $y < 2x - 1$
   $y \geq -2x + 3$

34. A spinner has the colors red, yellow and green on its three congruent sectors. What is the probability of the spinner landing on a yellow sector and then rolling a 4 on a die?

   (A) 1

   (B) $\dfrac{1}{2}$

   (C) $\dfrac{1}{18}$

   (D) $\dfrac{1}{36}$

35. A sequence is called **arithmetic** if the difference between successive terms is constant. Which of the following is *not* an arithmetic sequence?

   (A) 4, 7, 10, 13, ….

   (B) 0.08, 0.12, 1.6, 2, ….

   (C) $\dfrac{1}{3}, \dfrac{2}{3}, 1, \dfrac{4}{3}, ….$

   (D) 40, 40.6, 41.2, 41.8, ….

36. What is the simplest form of the quotient $\dfrac{\sqrt{96}}{\sqrt{6}}$?

   (A) $\sqrt{32}$

   (B) $4\sqrt{6}$

   (C) $12\sqrt{8}$

   (D) 4

GO ON

37. What is the perimeter of the triangle shown below?

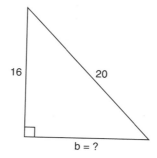

(A)   48

(B)   96

(C)   288

(D)   320

38. Which algebraic expression is a direct translation of the following?

The quantity of seven less b squared.

(A) $(b - 7)^2$

(B) $(7 - b)^2$

(C) $7b^2$

(D) $(7b)^2$

39. What is the slope of the graph that is represented by the following input-output table?

Input	Output
0	−2
3	0
6	2
9	4

(A) $\dfrac{-3}{2}$

(B) $\dfrac{-2}{3}$

(C) $\dfrac{2}{3}$

(D) $\dfrac{3}{2}$

40. What is the slope of the line whose equation is $6x + 3y = 18$?

(A) −3

(B) −2

(C) 2

(D) 3

41. What is the equation for the line of best fit in the scatter plot shown below?

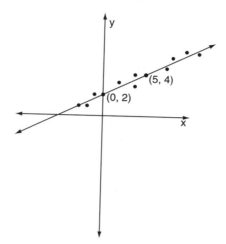

(A) $y = \dfrac{5}{2}x + 2$

(B) $y = \dfrac{-5}{2}x + 2$

(C) $y = \dfrac{-2}{5}x + 2$

(D) $y = \dfrac{2}{5}x + 2$

42. For which of the following inequalities would the graph consist of an open circle and an arrow to the right of -3?

(A) $-4x > 12$

(B) $x - 5 \geq -8$

(C) $x - 7 > -10$

(D) $-6x \leq -18$

GO ON

43. Using mental math, how can a student most easily calculate the value of $(15)(9)$?

    (A) $(15)(10 - 1)$

    (B) $(4 + 11)(7 + 2)$

    (C) $(20 - 5)(5 + 4)$

    (D) $(5 \times 3)(6 + 3)$

44. What is the simplest form for the value of $\sqrt{169} - \sqrt{81}$?

    (A) $\sqrt{88}$

    (B) 44

    (C) 4

    (D) 2

45. Look at the triangle shown below.

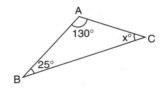

What is the value of x?

    (A) 45

    (B) 25

    (C) 15

    (D) 5

GO ON

46. What is the solution for n in the inequality $-2n + 9 \leq -11$?

    (A) $n \geq 10$

    (B) $n \geq -10$

    (C) $n \leq 10$

    (D) $n \leq -10$

47. Any set of inputs and outputs is called a _____.

    (A) function

    (B) domain

    (C) range

    (D) relation

48. To change the equation $\dfrac{2}{3}x + 4y = \dfrac{1}{6}$ to standard form, one must _____.

    (A) multiply the equation by 3

    (B) multiply the equation by 6

    (C) divide the equation by 3

    (D) divide the equation by 6

49. If $R = \{5, 6, 7, 9, 11, 13\}$ and $Q = \{4, 5, 6, 8, 10\}$, which set is represented by $R \cup Q$?

    (A) $\{5, 6\}$

    (B) $\{4, 5, 6, 8, 10\}$

    (C) $\{4, 5, 6, 7, 8, 9, 10, 11, 13\}$

    (D) $\{4, 7, 8, 9, 10, 11, 13\}$

GO ON

50. In a certain right triangle, the square of the hypotenuse is 225. If the square of one of the legs is 81, what is the length of the other leg?

    (A) 12

    (B) 20

    (C) 80

    (D) 144

51. Which one of the following is a rational number?

    (A) π

    (B) $\dfrac{7}{5}$

    (C) $\sqrt{101}$

    (D) 7.56239018....

52. In the diagram shown below, $\angle A \cong$?

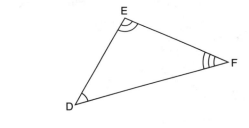

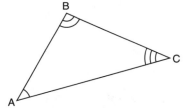

    (A) $\overline{AB}$

    (B) $\angle F$

    (C) $\overline{DE}$

    (D) $\angle D$

53. What is (are) the value(s) of n in the equation $|n| - 12 = -4$?

    (A) 8

    (B) -8

    (C) 16

    (D) 8 or -8

54. Which of the following best describes the following graph?

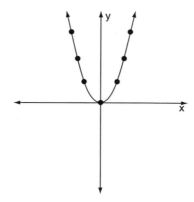

    (A) Linear relation

    (B) Perpendicular

    (C) Non-linear function

    (D) Linear function

GO ON

55. A bicycle rental office charges a $4 service fee, and then charges $2 for every hour that a bicycle is rented. If x represents the number of hours and y represents the total cost, which linear function could be used to model the fee schedule?

(A) $y = 2x + 4$

(B) $y = 4x + 2$

(C) $x = 2y + 4$

(D) $x = 4y + 2$

56. Which of the following equations has no solution?

(A) $|x - 5| + 2 = 4$

(B) $|x| - 5 = -2$

(C) $|x - 2| + 5 = 4$

(D) $|x| - 2 = 5$

57. Which of the following measures has the highest value for the following set of numbers?

12, 12, 14, 18, 974

(A) The median

(B) The mode

(C) The mean

(D) The second highest number

58. What is the value of $(5^2 + 2^5) \div 3^1$

    (A)   16

    (B)   18

    (C)   19

    (D)   21

59. Which of the following describes the meaning of $\overrightarrow{AB}$?

    (A)   Line AB

    (B)   Ray AB

    (C)   Plane AB

    (D)   Line segment AB

60. Given the formula $C = 2\pi r$, which expression is equivalent to π?

    (A) $\dfrac{C}{2r}$

    (B) $2rC$

    (C) $\dfrac{r}{2C}$

    (D) $C - 2r$

GO ON

61. Consider the following system of equations:

    $3x - y = 9$

    $5x + y = 7$

    What is the sum of the solutions for x and y?

    (A) 5

    (B) 1

    (C) −1

    (D) −5

62. Let n represent any number. Which of the following represents an interpretation of the expression $10 - \dfrac{n}{3}$?

    (A) Ten decreased by one-third of a number

    (B) One-third of a number subtracted by ten

    (C) One-third of the quantity ten less than a number

    (D) Ten divided by the difference of a number and three.

63. How many perfect square integers are there between 60 and 120?

    (A) 5

    (B) 4

    (C) 3

    (D) 2

GO ON

64. What is the value of $\dfrac{\sqrt{36}+2\sqrt{49}}{\sqrt{100}}$?

    (A) 1.34

    (B) 2

    (C) 2.6

    (D) 10

65. Apples cost \$0.40 a piece. The cost of five apples and four tomatoes cost less than \$4.00. Which inequality could be used to find the value of t, the cost of one tomato?

    (A) $(5)(0.40) - 4t < 4$

    (B) $\dfrac{0.40}{5} + \dfrac{t}{4} < 4$

    (C) $(5)(0.40) + 4t < 4$

    (D) $\dfrac{0.40}{5} - \dfrac{t}{4} < 4$

66. What is the value of $(2.1 \times 10^{20})(1.2 \times 10^{10})$?

    (A) 2.52×10^{30}

    (B) 3.3×10^{30}

    (C) 2.52×10^{200}

    (D) 3.3×10^{200}

GO ON

67. On a particular day, the lowest temperature was 8° below zero. By the afternoon, the temperature had risen by 20°. What was the temperature in the afternoon?

(A) 28°

(B) 14°

(C) 12°

(D) 6°

68. Which of the following equations represents a line whose slope is $\frac{1}{2}$ *and* whose y-intercept is -5?

(A) $2x - y = 5$

(B) $x - 2y = 10$

(C) $x - 2y = -10$

(D) $2x - y = -5$

69. Given the expression $y = ax - z$, which of the following is equivalent to z?

(A) $\dfrac{y}{ax}$

(B) $\dfrac{ax}{y}$

(C) $y - ax$

(D) $ax - y$

GO ON

70. For which of the following equations is zero the *only* solution?

(A) $|x - 6| + 5 = 5$

(B) $|x| - 10 = -10$

(C) $|x - 5| + 5 = 10$

(D) $|x| - 6 = 6$

Answer Key—Practice Test 2

Item Number	Correct Answer	Domain Description	Standard Measured
01	D	Numbers and Operations	M8N1
02	A	Geometry	M8G1
03	B	Algebra	M8A1
04	C	Algebra	M8A3
05	C	Data Analysis and Probability	M8D2
06	A	Algebra	M8A5
07	A	Data Analysis and Probability	M8D4
08	C	Data Analysis and Probability	M8P5
09	D	Numbers and Operations	M8N1
10	B	Geometry	M8G1
11	A	Algebra	M8A1
12	D	Algebra	M8A3
13	B	Algebra	M8A4
14	C	Data Analysis and Probability	M8D1
15	A	Algebra	M8A4
16	B	Data Analysis and Probability	M8P5
17	C	Algebra	M8P2
18	D	Numbers and Operations	M8N1
19	A	Geometry	M8G2
20	B	Algebra	M8A1
21	D	Algebra	M8A2
22	B	Algebra	M8A4
23	C	Algebra	M8A4
24	C	Algebra	M8A2
25	A	Data Analysis and Probability	M8D1

26	B	Data Analysis and Probability	M8P5
27	B	Numbers and Operations	M8N1
28	D	Numbers and Operations	M8N1
29	A	Geometry	M8G1
30	B	Algebra	M8A2
31	C	Algebra	M8A1
32	C	Algebra	M8A4
33	A	Algebra	M8A5
34	C	Data Analysis and Probability	M8D3
35	B	Algebra	M8A3
36	D	Numbers and Operations	M8N1
37	A	Geometry	M8G2
38	B	Algebra	M8A1
39	C	Algebra	M8A3
40	B	Algebra	M8A4
41	D	Data Analysis and Probability	M8D4
42	C	Algebra	M8A2
43	A	Numbers and Operations	M8N1
44	C	Numbers and Operations	M8N1
45	B	Geometry	M8G1
46	A	Algebra	M8A2
47	D	Algebra	M8A3
48	B	Algebra	M8A4
49	C	Data Analysis and Probability	M8D1
50	A	Geometry	M8G2
51	B	Numbers and Operations	M8N1
52	D	Geometry	M8G1
53	D	Algebra	M8A1
54	C	Algebra	M8A3
55	A	Algebra	M8A4
56	C	Algebra	M8A1

57	C	Data Analysis and Probability	M8P5
58	C	Numbers and Operations	M8N1
59	B	Geometry	M8G1
60	A	Algebra	M8A1
61	C	Algebra	M8A5
62	A	Algebra	M8A1
63	C	Numbers and Operations	M8N1
64	B	Numbers and Operations	M8N1
65	C	Algebra	M8A2
66	A	Numbers and Operations	M8N1
67	C	Numbers and Operations	M8N1
68	B	Algebra	M8A4
69	D	Algebra	M8A1
70	B	Algebra	M8A1

Practice Test 2—Progress Chart— Georgia CRCT

Numbers and Operations ___/ 14

01	09	18	27	28	36	43	44	51	58

63	64	66	67

Geometry ___/ 9

02	10	19	29	37	45	50	52	59

Algebra ___/ 36

03	04	06	11	12	13	15	17	20	21

22	23	24	30	31	32	33	35	38	39

40	42	46	47	48	53	54	55	56	60

61	62	65	68	69	70

Data Analysis and Probability ___/ 11

05	07	08	14	16	25	26	34	41

49	57

Total ___/ 70

Detailed Solutions

1. (D) Multiply the radicals and simplify. Thus, $\sqrt{80} \times \sqrt{5} = \sqrt{400} = 20$.

2. (A) By definition, parallel lines never intersect, as shown in the diagram below.

3. (B) The words "difference between" mean "subtraction." The "difference between n and twelve" is written as $n - 12$.

4. (C) If any vertical line intersects a graph at most once, then the graph expresses a function.

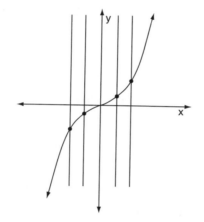

5. (C) Multiply all the different options.

   (3 sandwiches) $\times$ (3 side orders) $\times$ (5 beverages) = 45 combinations.

6. (A) Substitute $2x - 1$ for y in the second equation. Then, here are the steps.

   $$2x + 3(2x - 1) = 21$$
   $$2x + 6x - 3 = 21$$
   $$8x - 3 = 21$$
   $$8x - 3 + 3 = 21 + 3$$
   $$8x = 24$$
   $$\frac{8x}{8} = \frac{24}{8}$$
   $$x = 3$$

Now substitute 3 for x in either equation. Using the first equation, $y = 2(3) - 1 = 5$

7. (A) By definition, a line that most closely passes through the points of a scatter plot is called a line of best fit. (Technically, this line need not actually pass through any of the given points.)

8. (C) In a group of numbers, the mode is the number that appears most frequently. The number 9 occurs three times.

9. (D) 1,234,000 is also 1,234,000.0. Move the decimal point six places to the left so that the number is greater than or equal to 1 and less than ten. The result of this move is 1.234. Next, multiply 1.234 by 10^6.

10. (B) $\angle 1$ and $\angle 8$ are alternate exterior angles. Alternate exterior angles are congruent when lines are parallel. Then,

$$5x = 80$$
$$\frac{5x}{5} = \frac{80}{5}$$
$$x = 16$$

11. (A) Here are the steps.

$$2br = 7mn$$
$$\frac{2br}{2b} = \frac{7mn}{2b}$$
$$r = \frac{7mn}{2b}$$

12. (D) A relation is a function if each input has exactly one output. In the table, the input 7 has outputs 6 and 4.

13. (B) Isolate y to change the equation to slope-intercept form, as follows:

$$4x + 3y = 12$$
$$4x - 4x + 3y = -4x + 12$$
$$3y = -4x + 12$$
$$\frac{3y}{3} = \frac{-4x + 12}{3}$$
$$y = \frac{-4x}{3} + 4$$

Now, using the formula $y = mx + b$, $m = \frac{-4}{3}$.

14. (C) $\overline{A}$ means the complement of A. The complement is the set that contains everything in the universal set that is not contained in A.

15. (A) Use $y = mx + b$ to model each boy's total amount of money earned. Since $10 is a fixed donation, let it be b, the y-intercept. The money given for miles is $2 per mile, so 2 will be the slope. Inserting 2 and 10 in the correct places, we arrive at $y = 2x + 10$.

16. (B) The median is the number in the middle. Since 9 and 10 are both in the middle, find their mean which is $\dfrac{9+10}{2} = 9.5$

17. (C) The reason in step 2 suggests we had completed a step in division.

Therefore, the step must have shown the following: $\dfrac{7x}{7} = \dfrac{98}{7}$.

Simplifying, we arrive at $x = 14$.

18. (D) $\sqrt{147} = \sqrt{49} \times \sqrt{3} = 7\sqrt{3}$.

19. (A) Use the Pythagorean theorem because the figure is a right triangle.

$a^2 + b^2 = c^2$

$8^2 + b^2 = 17^2$

$64 + b^2 = 289$

$64 - 64 + b^2 = 289 - 64$

$b^2 = 225$

$\sqrt{b^2} = \sqrt{225}$

$b = 15$

20. (B) Here are the steps.

$x - (-52) = -43$

$x + 52 = -43$

$x + 52 - 52 = -43 - 52$

$x = -95$

21. (D) The correct graph shows all the values less than -2, but not equal to -2. The open circle at -2 implies that -2 is not part of the graph.

22. (B) Slope is defined as $\dfrac{rise}{run}$. As we proceed from $(0, 0)$ to $(1, -3)$ the rise is -3 and the run is 1. This ratio is also true in proceeding from $(1, -3)$ $(2, -6)$.

23. (C) The slope–intercept form of a line is $y = mx + b$, where m is the slope and b is the y-intercept. In the equation $y = -5x + 4$, the y-intercept is 4.

24. (C) Test the point (0, 0) to verify that the answer is $y \geq 2x - 1$

$$0 \geq 2(0) - 1$$
$$0 \geq -1$$

The line in the graph is solid, implying that it is part of the inequality. Thus the correct answer is $y \geq 2x - 1$

25. (A) There are no elements that occur in both P and R, so the solution is $\varnothing$, the empty set.

26. (B) The mean of the numbers is 5, since $\dfrac{1+3+5+7+9}{5} = \dfrac{25}{5} = 5$
Also, the median, the number in the middle, is 5.

27. (B) We note that $\sqrt{121} < \sqrt{129} < \sqrt{144}$. Since $\sqrt{121} = 11$ and $\sqrt{144} = 12$, we conclude that $11 < \sqrt{129} < 12$.

28. (D) $(-2)(-2)(-2)(-2)(-2) = -32$.

29. (A) By definition, angles 4 and 8 are corresponding angles.

30. (B) Sammy can spend $40 for his sneakers. He can also spend any amount less than $40.

31. (C) Substitute 774 for y in the equation $y = 18x$. Then $774 = 18x$. Thus, $\dfrac{774}{18} = \dfrac{18x}{18}$, so $x = 43$.

32. (C) Let $y = 0$ to find the x-intercept. Here are the steps.

$$6x + 4(0) = 12$$
$$6x = 12$$
$$\dfrac{6x}{6} = \dfrac{12}{6}$$
$$x = 2$$

The x-intercept is (2, 0)

33. (A) Test the inequalities using (0,0). Begin with the inequality $y > 2x - 1$. We find that $0 > 2(0) - 1$. This leads to $0 > -1$, which is true. In addition, we must substitute (0, 0) into the inequality $y \leq -2x + 3$. Then, $0 \leq -2(0) + 3$. This leads to $0 < 3$, which is also true. Therefore, answer choice (A) is correct.

34. (C) The location of the spinner and the roll of a die are independent events, so the probabilities are multiplied. The probability of the spinner landing in a yellow sector is $\dfrac{1}{3}$. The probability of rolling a 6 is $\dfrac{1}{6}$. Thus, the required probability is $\dfrac{1}{3} \times \dfrac{1}{6} = \dfrac{1}{18}$.

35. (B) The difference between the first two terms is $0.012 - 0.08 = 0.04$, but the difference between the second and third terms is $1.6 - 0.12 = 1.48$. Since these differences are not the same, the sequence is not arithmetic. Notice that each of answer choices (A), (C), and (D) has a common difference between successive terms.

36. (D) $\dfrac{\sqrt{96}}{\sqrt{6}} = \sqrt{16} = 4.$

37. (A) Perimeter is the distance around the figure. Use the Pythagorean theorem to find the missing side.

$16^2 + b^2 = 20^2$, followed by $256 + b^2 = 400$.

Then $256 - 256 + b^2 = 400 - 256$, which leads to $b^2 = 144$. This means that $b = 12$. Finally, $12 + 16 + 20 = 48$.

38. (B) "The quantity" implies parentheses. "Seven less b" means $7 - b$.

Then $7 - b$ is to be squared, which means that the answer is $(7 - b)^2$.

39. (C) Starting from $(0, -2)$, a rise of 2 units and a run of 3 units leads to the point $(3, 0)$. From this point, a rise 2 units and a run of 3 units leads to the point $(6, 2)$. Thus, $\dfrac{rise}{run} = \dfrac{2}{3}.$

40. (B) Isolate y to transform the equation to slope-intercept form.

$$6x + 3y = 18$$
$$6x - 6x + 3y = -6x + 18$$
$$3y = -6x + 18$$
$$\frac{3y}{3} = \frac{-6x}{3} + \frac{18}{3}$$
$$y = -2x + 6$$

Thus, the slope, m, is -2.

41. (D) Find two points that lie on or close to the line, namely $(0, 2)$ and $(5, 4)$. From $(0, 2)$ to $(5, 4)$, the rise is 2 units and the run is 5 units. The line intersects the y-axis at $(0, 2)$. Therefore, b, the y-intercept, is 2. Thus, the required equation is $y = \dfrac{2}{5}x + 2$.

42. (C) To solve the inequality $x - 7 > -10$, add 7 to both sides. Then the inequality becomes $x > -3$. Its graph is an open circle on -3 and an arrow to the right. The graph for answer choice (A) is an open circle on -3 and an arrow to the left, since the inequality changes when we divide by a negative number. The graph for answer choice (B) is a closed circle on -3 and an arrow to the right. The graph for answer choice (D) is a closed circle on 3 and an arrow to the right.

43. (A) $(15)(9)$ can be thought of as $(15)(10 - 1) = 150 - 15 = 135.$

44. (C) $\sqrt{169} - \sqrt{81} = 13 - 9 = 4$

45. (B) The sum of the measures of the angles in a triangle is 180°. Add 130 and 25 and subtract the sum from 180. Thus, $180 - (130 + 25) = 25$.

46. (A) Here are the steps.

$$-2n + 9 \leq -11$$
$$-2n + 9 - 9 \leq -11 - 9$$
$$-2n \leq -20$$
$$\frac{-2n}{-2} \geq \frac{-10}{-2}$$
$$n \geq 10$$

Remember to reverse the direction of the inequality sign when multiplying or dividing by a negative number.

47. (D) All functions are relations, but not all relations are functions. Any set of inputs and outputs is a relation.

48. (B) The required step is to multiply the given equation by 6, in order to eliminate any fractions.

Then $6\left(\frac{2}{3}x + 4y = \frac{1}{6}\right)$

$$4x + 24y = 1$$

49. (C) $R \cup Q = R + Q - (R \cap Q)$

Thus, $\{5, 6, 7, 9, 11, 13\} + \{4, 5, 6, 8, 10\} - \{5, 6\} = \{4, 5, 6, 7, 8, 9, 10, 11, 13\}$.

50. (A) Let b represent the length of the missing leg. Using the Pythagorean theorem, $81 + b^2 = 225$. Then $b^2 = 225 - 81 = 144$. Thus, $b = \sqrt{144} = 12$.

51. (B) The decimal portion of a rational number either terminates or contains a repeating pattern. $\frac{7}{5} = 1.4$, which represents a terminating decimal. All the other selections do not terminate nor do they contain a repeating pattern.

52. (D) $\angle A$ and $\angle D$ both display single arcs denoting that their measures are the same. Angles with equal measures are congruent.

53. (D) Here are the steps to the solution.

$$|n| - 12 = -4$$
$$|n| - 12 + 12 = -4 + 12$$
$$|n| = 8$$

Therefore, $n = 8$ or $n = -8$.

54. (C) The graph does not lie along a straight line, so it is not a linear function. It does pass the vertical line test so it is a non-linear function.

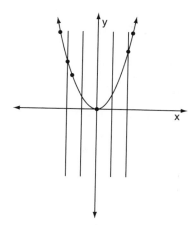

55. (A) The slope-intercept form of a line, $y = mx + b$, can be used to model the fee schedule. Since the $4 service charge is fixed, this is the b value (y-intercept). The rate varies depending on the hours rented. Therefore the hourly fee, $2, is the slope. This is the value of m.

Thus, the correct equation is $y = 2x + 4$.

56. (C) The first step to solving the equation $|x-2|+5=4$ is to subtract 5 from each side. Then the equation becomes $|x-2|=-1$. But the absolute value of $x - 2$ must be at least zero. Thus, this equation cannot have a solution. Each of answer choices (A), (B), and (D) has two answers. For answer choice (A), the answers are 3 and 7; for answer choice (B), the answers are 3 and -3; for answer choice (D), the answers are 7 and -7.

57. (C) To find the mean, add up all the numbers and divide by 5. Then the mean is $\frac{12+12+14+18+974}{5} = \frac{1030}{5} = 206$. Note that the mean is considerably larger than all the numbers except 974. The values of answer choices (A), (B), and (D) are 14, 12, and 18, respectively.

58. (C) $(5^2 + 2^5) \div 3^1 = (25 + 32) \div 3 = 57 \div 3 = 19$.

59. (B) The symbol "$\rightarrow$" above two letters refers to a ray.

60. (A) Here are the steps.

$$C = 2\pi r$$
$$\frac{C}{2r} = \frac{2\pi r}{2r}$$
$$\frac{C}{2r} = \pi$$

61. (C) Adding the equations yields $8x = 16$, from which we determine that $x = 2$. Substitute this value of x into either equation. Using the first equation, $(3)(2) - y = 9$, which means $6 - y = 9$. Subtracting 6 from each side results in $-y = 3$. So $y = -3$. Therefore the sum of the values of x and y is $2 + (-3) = -1$.

62. (A) The expression $\frac{n}{3}$ means "one-third of a number." When the number 10 is followed by a minus sign, it means "ten decreased by."

63. (C) The lowest perfect square above 60 is 64, since $8^2 = 64$. We note that $9^2 = 81$ and $10^2 = 100$. Since $11^2 = 121$ which is more than 120, the only three perfect squares that satisfy the given conditions are 64, 81, and 100.

64. (B) $\dfrac{\sqrt{36} + 2\sqrt{49}}{\sqrt{100}} = \dfrac{6 + (2)(7)}{10} = \dfrac{20}{10} = 2.$

65. (C) The cost of five apples is represented by $(5)(0.40)$. The cost of t tomatoes is represented by $4t$. Since their sum is less than \$4.00, we can write the inequality (without the dollars and cents symbols) as $(5)(0.40) + 4t < 4$.

66. (A) When multiplying numbers in scientific notation, we multiply the decimal parts that lie between 1 and 10, then add the powers of 10. $(2.1)(1.2) = 2.52$ and $10^{20} \times 10^{10} = 10^{20+10} = 10^{30}$.

67. (C) The morning temperature could be written as $-8°$. Since the temperature rose by $20°$, the new temperature was $-8° + 20° = 12°$.

68. (B) In slope-intercept form, the equation would be $y = \frac{1}{2}x - 5$. Multiply both sides of this equation by 2 to get $2y = x - 10$. For both sides of this equation, subtract $2y$ and add 10. The result will be $x - 2y = 10$. Answer choice (A) is wrong because the slope would be 2. Answer choice (C) is wrong because the y-intercept would be 5. Answer choice (D) is wrong because the slope would be 2 and the y-intercept would be 5.

69. (D) First subtract ax from each side to get $y - ax = -z$. Since $-z = -1z$, we must divide both sides of the equation by -1. So $z = -y + ax$, which is equivalent to $ax - y$.

70. (B) Adding 10 to each side of $|x| - 10 = -10$ yields $|x| = 0$. The only possible answer is 0. For answer choice (A), the solution is 6. For answer choice (C), the solutions are 0 and 10. For answer choice (D), the solutions are -12 and 12.

INSTALLING REA's TestWare®

SYSTEM REQUIREMENTS

Pentium 75 MHz (300 MHz recommended), or a higher or compatible processor; Microsoft Windows 98 or later; 64 MB available RAM; Internet Explorer 5.5 or higher.

INSTALLATION

1. Insert the Georgia CRCT Grade 8 Math TestWare® CD-ROM into the CD-ROM drive.
2. If the installation doesn't begin automatically, from the Start Menu choose the RUN command. When the RUN dialog box appears, type d:\setup (where *d* is the letter of your CD-ROM drive) at the prompt and click OK.
3. The installation process will begin. A dialog box proposing the directory "Program Files\REA\GACRCT_MATH" will appear. If the name and location are suitable, click OK. If you wish to specify a different name or location, type it in and click OK.
4. Start the Georgia CRCT Grade 8 Math TestWare® application by double-clicking on the icon.

REA's Georgia Grade 8 Math TestWare® is **EASY** to **LEARN AND USE**. To achieve maximum benefits, we recommend that you take a few minutes to go through the on-screen tutorial on your computer. The "screen buttons" are also explained here to familiarize you with the program.

TECHNICAL SUPPORT

REA's TestWare® is backed by customer and technical suppor. For questions about **installation or operation of your software**, contact us at:

Research & Education Association
Phone: (732) 819-8880 (9 a.m. to 5 p.m. ET, Monday–Friday)
Fax: (732) 819-8808
Website: www.rea.com
E-mail: info@rea.com

Note to Windows XP Users: In order for the TestWare® to function properly, please install and run the application under the same computer administrator-level user account. Installing the TestWare® as one user and running it as another could cause file-access path conflicts.